MEDICAL LAW AND MORAL RIGHTS

MEDICAL LAW AND MORAL RIGHTS

by

CARL WELLMAN
Washington University, St.Louis, USA

Springer

A C.I.P. Catalogue record for this book is available from the Library of Congress.

ISBN-10 1-4020-3751-1 (HB)
ISBN-13 978-1-4020-3751-1 (HB)
ISBN-10 1-4020-3752-X (e-book)
ISBN-13 978-1-4020-3752-8 (e-book)

Published by Springer,
P.O. Box 17, 3300 AA Dordrecht, The Netherlands.

www.springeronline.com

Printed on acid-free paper

Printed in the Netherlands.

CONTENTS

DETAILED TABLE OF CONTENTS

6. The Concept of Fetal Rights

This essay reports the introduction of the right of the unborn child not to be injured and explains why the legal rights of the fetus were traditionally understood to be conditional on birth. It then examines the ways in which the courts have struggled to determine precisely when the fetus becomes a separate individual from its mother and to explain how the fetus can have any rights before birth if they can be claimed only after birth. It argues that it is misleading to ascribe rights to the fetus but that this might be a useful legal doctrine based on the legal fiction of fetal agency.

7. Maternal Duties and Fetal Rights

This essay explains the grounds of the moral duties of the mother to care for and not abuse her fetus. It examines critically the arguments of those who advocate giving the fetus legal rights holding against its mother in order to enforce these moral duties. After considering objections to this proposal, it recommends giving the fetus very limited legal rights against the mother.

8. The Scope of the Right to Procreational Autonomy

This essay examines a number of judicial opinions, mostly in cases before the United States Supreme Court, in order to define the content of the existing constitutional right to procreational autonomy of married persons. It argues that there is an analogous human right to procreational autonomy. It concludes that the constitutional right ought to be somewhat broader in scope than this underlying moral right.

9. Possessors of the Right to Procreational Autonomy

This essay examines a variety of judicial opinions in order to identify the kinds of individuals who now possess the constitutional right to procreational autonomy. It then appeals to the human rights to procreational autonomy, privacy, personal security, and equitable treatment to argue that even persons living in same-sex relationships, minors, mentally retarded persons, child abusers, and carriers of serious genetic defects ought to possess this right.

10. Medical Futility: Legal Duties and Moral Rights

This essay reports recent court cases in which physicians and hospitals have sought to deny medical treatment demanded by their patients on the grounds that it would be medically futile. It examines critically both the reasons for giving physicians a legal liberty to refuse to provide futile medical treatment even when demanded by their patients and the reasons against giving physicians any such legal liberty. It concludes in favor of establishing this legal liberty.

CHAPTER 1

INTRODUCTION

This is a collection of essays written over the past 5 years.[1] Although each essay can be read and understood independently of the others, the volume as a whole reveals several unifying themes and purposes. All of the essays deal directly or indirectly with unresolved legal issues in contemporary medical law. Several of them explicitly advocate specific reforms of the existing United States law. Although the examined court cases are almost entirely those within my own country, the legal issues they illustrate are of international interest. They arise from recent developments in medical practice that are spreading from country to country along with the globalization of modern medical technology. These legal issues interest me as a moral philosopher because they concern moral problems that arise in medical practice. Hence, several of the essays propose specific solutions to some of the most urgent moral problems that confront potential parents, patients, physicians, or hospital administrators making difficult moral choices. These conclusions are based upon mid-level moral theory. Rather than explore the most general and abstract ethical theory, most of the essays define the content and identify the grounds of several fundamental moral duties and moral rights relevant to medical law and medical practice. Finally, all of the essays presuppose, to a greater or lesser extent, the theory of rights I have developed during more than three decades. Hence, they should serve to test the usefulness, or lack of it, of this abstract theory of rights when it is applied to live issues in medical law and controversial moral choices in medical practice.

1.1. Legal Issues

The following essays concern important questions about how our courts and legislatures ought to resolve currently undecided issues in medical law. Ought there to be a legal right to physician-assisted suicide? Ought fetuses to have legal rights holding against their mothers? What ought to be the scope of the constitutional right to procreation and who ought to possess this right? Ought physicians and hospital administrators to have the legal right to refuse to provide futile medical treatment demanded by their patients? These are all normative questions, questions about what the law ought to be, not what it actually is.

Why, then, do I approach these questions via an examination of recent court cases, cases in which the judges have decided between the conflicting claims of the parties as to their rights under existing law? One reason is to ensure that these essays focus on open

[1] The three essays that have been previously published have been slightly revised to take account of more recent literature and to conform to the format of this volume.

questions, real legal issues that need to be resolved in contemporary medical law. Another is to identify the factual circumstances, facts specified in the published judicial opinions, under which these questions arise and that are relevant to their resolution. Finally, the opinions of the courts together with any concurring and dissenting opinions point to most of the pro and con arguments that one should consider before advocating any moral reform of the existing law.

"*Glucksberg* v. *Compassion*" argues that the issue of whether the Due Process Clause grounds a constitutional right to physician-assisted suicide remains open in spite of the decision of the United States Supreme Court in *Washington* v. *Glucksberg*. However, "A Legal Right to Physician-Assisted Suicide" advocates the introduction by the several state legislatures of a statutory right to physician-assisted suicide. It bases this recommendation upon the need to spare qualified patients unnecessary suffering, to enable them to die with dignity, and to respect their autonomy. These are, of course, familiar reasons frequently discussed in the legal and moral literature. What is original in this essay, and in the essay defending an analogous moral right, is the way in which these reasons are reformulated to make their import clearer as well as the explanations of the underlying reasons that give them moral force. It is worthy of note that one can appeal to respect for the patient's autonomy without presupposing any controversial human right to personal self-determination. This essay also defends its conclusion against several serious objections, especially that any legal right to physician-assisted suicide would be abused.

"Maternal Duties and Fetal Rights" advocates the modest expansion of old fetal rights and the introduction of new fetal rights holding against the mother and the father of the unborn child. It argues that this is necessary to prevent some children from being born suffering serious harms and that it would not be unfair to either parent because of their moral duties of parental care to the fetus. Original and important aspects of this essay are its explanation of the nature and grounds of the pregnant woman's moral duties to her unborn child and the limits imposed upon new fetal rights in order to meet objections fatal to more sweeping proposals for fetal rights. It also explains the relevance of moral maternal duties to legal reform without assuming that all or even most moral duties ought to be legally enforced.

"The Scope of the Right to Procreational Autonomy" argues that this right of married individuals ought to be defined very broadly. Specifically, their constitutional liberty-rights to procreate or not procreate and to prevent or not prevent procreation should be defined in very general terms limited only on independent grounds, such as the duties not to rape or cause a miscarriage by a physical attack upon a pregnant woman. The argument rests upon the human rights to procreational autonomy and to privacy, each explained in an original manner. However, the essay concludes that this constitutional right may justifiably be overridden when, but only when, this is necessary to protect some compelling state interest.

"Possessors of the Right to Procreational Autonomy" argues that unmarried adults in stable heterosexual relationships, adults in stable same-sex relationships, adults not living in any stable relationship with a partner, adults on welfare, persons who carry serious genetic defects, mentally retarded persons, minors, and even child abusers ought to possess the same right to procreational autonomy as married persons. However, the courts ought to be authorized to limit the exercise of this right by mentally retarded persons or minors under special circumstances. At various points, the arguments appeal to the human rights to procreational autonomy, to privacy, to equitable treatment, and to personal security. The explanations of these fundamental moral rights may be of more interest than the controversial conclusions drawn from them.

"Medical Futility: Legal Duties and Moral Rights" advocates the recognition of a legal liberty of the physician to refuse to provide futile medical treatment demanded by the patient or her family. However, the physician's individual judgment of medical futility should be subject to judicial review in light of the standards of good medical practice of the medical profession as a whole, and a physician who refuses to provide demanded medical treatment should have a legal duty to co-operate with the patient in finding another physician who will provide the treatment she desires. Moral rights enter into the argument not so much as reasons to justify its conclusion as considerations to be discounted as irrelevant or unavailing.

1.2. Moral Problems

The court cases reported in these essays are not evidence that our society is overly litigious. They reflect urgent moral problems that arise in contemporary medical practice. Patients, physicians, and hospital administrators face difficult choices and often wonder how they are morally permitted or required to act. Patients who believe that suicide is normally immoral wonder whether it would be morally permissible for them to shorten a terminal illness or escape intolerable suffering by committing suicide with the help of their physicians. Physicians who recognize their professional duty to preserve human life are uncertain how they ought to respond to desperate patients who request assistance in committing suicide. "A Moral Right to Physician-Assisted Suicide" argues that patients who are terminally ill or undergoing intolerable irremediable suffering do have the moral liberty-rights to request, obtain, and use assistance from their attending physicians to end their lives. This implies that it is normally permissible for them to act in any or all of these three ways. Although it does not imply that their physicians are morally required to assist them in committing suicide, it is permissible for their attending physicians to do so if they so choose. Physicians do, however, have a moral obligation not to prevent their patients from finding another physician who is willing to assist them in ending their lives. Thus, this essays does give patients and physicians guidance to assist them in making difficult and inescapable moral choices.

"Maternal Duties and Fetal Rights" argues that the maternal duties to care for and not to abuse one's child after birth also apply to the fetus. Its analysis of the scope of these moral obligations implies that pregnant women have a moral duty not to engage in conduct that would case any serious avoidable harm to their children and a moral duty to submit to fetal therapies needed to preserve the health of children to whom they will give birth, at least if to do so will not impose excessive suffering upon themselves. This conclusion informs them in general terms regarding some kinds of conduct that are respectively impermissible and required.

Pregnant women who have been informed that their unborn child is genetically defective often wonder whether it is morally permissible to have an abortion. A divorced man unwilling to father a child with the woman who was his wife may be uncertain whether it is permissible for him to refuse to consent to her use of fertilized ova they had preserved before their separation. A conscientious physician may well wish to know whether she ought to perform an abortion requested for some trivial reason. "The Scope of the Right to Procreational Autonomy" defines the content of the human rights to procreate and to prevent procreation. These imply that, at least for married individuals, it is generally permissible to procreate or refrain from procreation as one desires. However, this bilateral moral liberty is limited by the moral duties not to procreate without the consent of one's partner, not to knowingly risk the birth of a child so seriously defective that she would not have a life worth living, and not to procreate when one is unable or unwilling to rear or arrange for others to rear one's child. These duties also limit their human right to prevent or not prevent procreation. This essay also defines the human right to privacy implying that it is morally wrong for physicians and others to prevent or hinder married individuals from exercising their human procreational rights. Thus, this essay speaks to the morally contested choices that confront many individuals in our society today.

"Possessors of the Right to Procreational Autonomy" argues that unmarried individuals in stable heterosexual relationships or in stable same-sex relationships, adults not living in any stable relationship with a partner, recipients of welfare benefits, persons who carry serious genetic defects, child abusers, some mentally retarded persons, and some minors also possess the human rights to procreate and to prevent procreation. To be sure, the exercise of these rights by the last three categories of individuals is limited to a considerable extent by their moral duties not to procreate when this would risk the birth of a child so seriously defective that she would not have a life-worth living or when one is unable to rear or arrange for others to rear one's child. This essay also argues that the procreational decisions of these possessors are generally protected by their human right to privacy. Accordingly, it extends the moral advice of the previous essay far beyond the class of married individuals.

Physicians and hospital administrators usually recognize their moral obligation to respect the autonomy of their patients, but at the same time they believe that the provision of futile treatment is at least medically inappropriate and often morally objectionable. How, then, ought they to respond when a patient requests futile medical treatment? "Medical Futility: Legal Duties and Moral Rights" argues that the patient's moral rights not to be abandoned, to autonomy, to life, and to nondiscrimination do not impose any moral duty to provide futile medical care even when this is demanded by a patient. Hence, it implies that physicians are morally permitted, but not thereby morally required, to refuse to provide such medical treatment. Thus, several of these essays do respond reasonably, but not necessarily infallibly, to serious moral problems that arise in contemporary medical practice.

1.3. Moral Theory

On a more general but not the most abstract level, several of these essays make original contributions to moral theory, especially to the theory of human rights. They do this partly by defining the content of specific fundamental moral rights to which jurists and moral philosophers typically appeal in vague and indeterminate language. More important, and as one would expect more controversial, are their suggestions regarding the specific moral reasons that ground these moral rights.

"A Moral Right to Physician-Assisted Suicide" defines this right as a rights-package consisting of the three moral liberty-rights to request or not request, to obtain or not obtain, and to use or not use assistance from one's physician to commit suicide. Each of these constituent rights rests on its own grounds. (1) If one grants for the sake of the argument that there is, in general, a moral duty not to request assistance to commit suicide, then the moral liberty to request assistance is suspect and requires one or more liberty-conferring reasons. These are either that this is necessary for fulfilling a more stringent duty or to escape intolerable irremediable suffering. The liberty not to request assistance to use in committing suicide is an innocent liberty because there is under normal circumstances no moral duty to request such assistance. Hence, it is grounded on the absence of any contrary duty-imposing reason. (2) The liberty to obtain assistance one could use to commit suicide is an innocent liberty because to do so is neither to implicate one's physician in any immoral activity nor to put oneself in a position to violate a moral duty. The liberty not to obtain assistance is, in general, also an innocent liberty because there is no contrary moral duty except under very special circumstances. (3) The liberty to use assistance obtained from one's physician to commit suicide is a suspect liberty because there is a general moral duty not to kill oneself. However, this duty is overcome when committing suicide is necessary to fulfill a more stringent moral duty or when to continue enduring such suffering would be to endure an excessive sacrifice. The liberty not to use any assistance obtained from one's physician to commit suicide is generally an innocent liberty grounded on the absence under normal circumstances of any moral reason imposing a duty to kill oneself.

"The Scope of the Right to Procreational Autonomy" defines the human right to procreational autonomy as a rights-package consisting of the two moral liberty-rights to procreate or not procreate and to prevent or not prevent procreation. Again, each of these constituent rights has its own grounds. (1) The moral liberty to procreate is, in general, an innocent liberty because there is no general moral reason sufficient to impose a duty to refrain from procreation. It is not true that human desires inevitably render one's life miserable or that our planet is so overpopulated that one is typically wrong to bring another child into existence. However, one's liberty to procreate is limited by the special moral duties not to procreate without the consent of one's partner, not to knowingly risk giving birth to a child so seriously defective as to have a life not worth living, and not to procreate when one is unable or unwilling to rear or arrange for others to rear one's child. The moral liberty not to procreate is also, in general, an innocent liberty because under normal circumstances one does not have any moral duty to procreate. (2) Although the moral liberty to prevent procreation by abstaining from sexual intercourse or using contraceptives is an innocent liberty, the liberty to prevent procreation by having an abortion is a suspect liberty. This is because there is a general moral duty not to intentionally destroy a potential human life. However, this duty is overcome when the child would be born so defective as to have a life not worth living, or the parents would be unable to care for their child, or to have a child would impose an excessive sacrifice upon one or both parents. The moral liberty not to prevent procreation is, in general, an innocent liberty because under normal circumstances, there is no moral duty not to procreate.

This essay also defines the human right to the privacy of one's decisions. The core of this fundamental moral right is the claim against others that they do not prevent or hinder one from acting on any of one's private decisions. A private decision is one that affects the life of the agent for better or worse and such that no one else has a sufficient reason to prevent or hinder one from acting on it. It is a choice such that how one acts makes a difference to the value of one's own life and is no one else's business. Because a moral claim of X against Y consists in a correlative moral duty of Y to X together with the moral power of X to claim performance of this duty, its grounds consist of the grounds of Y's duty together with the grounds of X's power. There is a general *prima facie* moral duty not to prevent or hinder any human being from acting as she chooses. This is partly because any interference with one's action is experienced as disturbing and frustrating and normally resented to a degree that damages personal relationships. Interference also disrupts one's activities and often prevents or hinders one from pursuing projects of considerable importance to one's life. These harms are moral reasons for others not to interfere with one's decisions and for those in society with one who does interfere to react negatively to him. Although this general duty is sometimes negated by contrary moral reasons, this is by definition not so when the decision is a private one. The ground of X's moral power to claim performance of this duty rests upon the fact that the action of demanding or even requesting Y not to interfere together with pointing out that one has decided to act on a private decision changes the moral situation in a way that makes Y's duty of noninterference more stringent. This is because any subsequent interference is a flagrant disregard of the wishes of another and more than normally destructive of their personal relationship and will show oneself to be especially domineering and hostile to the claimant's projects.

"Possessors of the Right to Procreational Autonomy" defines the human right to equitable treatment in a way that makes clear what kinds of reasons are needed to establish its relevance or irrelevance to the case at hand. The human right to equitable treatment is a fundamental moral claim-right not to be treated less well than others who are similarly situated unless there is a justicizing reason to do so. Two or more persons are similarly situated when they have the same claim to the kind of treatment under consideration. This establishes their *prima facie* right to equal treatment. Nevertheless, it may not be inequitable to treat one of these persons less well if there is some just-making reason to do so. A just-making or justicizing reason is a special case of a justifying reason; it is one that explains why the unequal treatment is not unfair. Thus, two workers on an assembly line are similarly situated regarding payment because they both have a moral claim to be paid for their labor. Hence, their human right to equitable treatment requires equal pay for equal work. But a justicizing reason for paying Jim less than Jane would be that Jim has produced fewer widgets than Jane did last week.

"Medical Futility: Legal Duties and Moral Rights" proposes an interpretation of the human right to life that renders much more determinate this right so often asserted but so seldom defined. The human right to life should be understood to be a rights-package including at least the moral claim-right not to be killed by another, the moral claim-right that others not endanger one's life, the moral liberty-right to defend one's life with all necessary force, the moral liberty-right to preserve one's life by any necessary means, and perhaps the moral claim-right to be rescued from the danger of death. Most controversial is the suggestion that these are fundamental moral rights concerning life in the sense of *a life*, a conscious biography, and not merely biological existence as a living organism. One's biological functioning is relevant to one's human right to life only because it is a necessary condition for any human being to have a conscious biography. No doubt these contributions to the moral theory of human rights need considerably more explanation and justification, but they should be sufficient to stimulate fruitful philosophical debate. And their adequacy can be partly confirmed or disconfirmed by whether their applications in these essays yield acceptable or unacceptable moral conclusions.

1.4. A Conception of Rights

All of these essays presuppose to a greater or lesser extent a way of thinking about legal and moral rights explained and defended in *A Theory of Rights* and *Real Rights*. One of the purposes of applying this abstract theory to specific issues in medical law and biomedical ethics is to test its adequacy. Typically, the appeal to rights, especially human rights, occasions dogmatic assertions and counter assertions preventing any practical compromise or reasonable discussion. How, if at all, is this conception of rights useful in resolving controversial practical questions concerning particular cases?

Three aspects of this conception of legal rights structure the reasoning in "Defining the Rights to Physician-Assistant Suicide." First, any right is assumed to be defined by a single core legal position. Thus, the physician's right to give assistance to the patient is distinguished from the patient's right to receive and use assistance from her physician. This is a useful, even a necessary, preliminary to understanding how the physician's right depends upon and can be derived from the patient's constitutional right. Then, the patient's right is analyzed into three component rights and taken to be a rights-package rather than a single unanalyzed right. The resulting definition is much more complex than those, such as "a right to die" or "the liberty to shape death," proposed by Justice Reinhardt or even the allegedly more careful formulation, "a right to suicide which itself includes a right to assistance in doing so," substituted by Chief Justice Rehnquist. This would be useful to a patient contemplating physician-assisted suicide because it would make explicit the three distinct kinds of action permitted to her by the law.

Second, the modality of each core position is made explicit because rights are defined in terms of Hohfeld's fundamental legal conceptions. For example, the patient's legal right to physician-assisted suicide is conceived of as a rights-package defined by three bilateral legal liberties—the liberty to request or not request, the liberty to accept or not accept, and the liberty to use or not use assistance provided by one's physician to commit suicide. Specifying the modality of these rights as liberties rather than as claims reveals the fallacy in the reasoning of those who argue that granting the patient's right to physician-assisted suicide would thereby impose a duty to provide assistance upon the attending physician. Moreover, the limits of each of these liberties of action are drawn with some precision. This second aspect of the presupposed conception of rights is also useful to any patient contemplating physician-assisted suicide because it provides very clear guidance concerning what actions are and are not legally permissible. In addition, this aspect would be useful to any lawyer or physician advising the patient regarding her legal rights. For judges, this sort of analysis would be a mixed blessing. It would define with more precision the legal issues before any court deciding a case concerning physician-assisted suicide. This would help to identify the relevant legal sources and the kinds of reasoning necessary in any convincing judicial opinion. At the same time, it might determine the outcome of future cases more definitely than a court would want. In a common law jurisdiction like that of the United States, judges try to define the law incrementally, one small step at a time. Sweeping legal reforms are too dangerous because one cannot reliably predict all the relevant circumstances of cases that may face the courts in the future. Hence, the clarity and precision so prized by analytic philosophers proficient in testing a theory against imagined cases is less welcome to practicing judges who must decide actual cases with very real consequences for the parties before the court.

Third, any legal right is conceived of as a complex structure including associated legal positions that, if respected, give dominion over the defining core to the right holder in any potential confrontation with one or more second parties. For example, each of the core liberties in the patient's legal rights-package to physician-assisted suicide is protected by a legal immunity against legislation that would extinguish or unduly burden it. Making this explicit explains just how the physician's constitutional right to give assistance can be derived from the patient's right to receive assistance from the physician, not from its core liberties but from its protective constitutional immunities.

The description of the proposed right to physician-assisted suicide in "A Legal Right to Physician-Assisted Suicide" also presupposes the conception of a legal right as a complex structure of Hohfeldian legal positions. It spells out the important associated positions more fully than in any of the previous essays. This would provide guidance to any state legislator introducing a bill proposing a statute of the recommended kind and could help to focus debate within a legislature or in the wider forum of public opinion.

The most serious problem with the language of rights is its obscurity. One typically refers to a right by using a mere label or a compressed description, for example, the right to "life" or the right "not to be killed unjustly." This causes misunderstanding because different speakers often interpret rights differently and renders the practical implications of any right indeterminate and controversial. "A Moral Right to Physician-Assisted Suicide" illustrates how this right can best be interpreted by using the author's conception of rights. First, it analyzes this right into a rights-package consisting of three moral liberty-rights. This makes it clear that three distinct moral issues are involved in debates about this right: whether suicide is ever morally permissible, whether requesting assistance is morally permissible, and whether obtaining the means to commit suicide is permissible. This threefold focus should help to resolve personal uncertainty and interpersonal disagreement by showing how three sets of moral reasons are required to settle these issues. Second, this essay presupposes that any genuine right must include associated moral positions sufficient to confer dominion upon the right holder. Hence, it recognizes that each of these defining core liberties is protected by at least one or more duties of noninterference. Although, the presupposed conception of rights does not identify these duties, much less their grounds, it does contribute to fruitful thinking by suggesting the direction in which one must look to find any adequate solution to the moral problems posed by physician-assisted suicide.

"The Concept of Fetal Rights" describes several of the conceptual problems our courts have faced in deciding cases concerning the legal rights of unborn children. It argues that ascribing rights to the unborn is misleading because this describes wrongly the real legal situation. On the best conception of rights, fetuses are not possible right holders. The real right holders are the parents or guardians who exercise their own rights regarding the child, normally after birth. This essay thereby contributes to jurisprudence by providing a more accurate way to interpret these asserted legal rights. It does not, however, infer that all ascriptions of fetal rights are either meaningless or mistaken. Although they do presuppose a legal fiction, legal fictions are often useful in formulating legal doctrines and deciding hard cases before the courts. It even suggests options the courts have hesitated to adopt in the past, such as abandoning the born alive rule. In this way, it could be useful to judges deciding particular cases as well as philosophers of law concerned with more general legal theory.

"The Scope of the Right to Procreational Autonomy" presupposes the same conception of rights in its treatment of human rights, especially the human right to privacy.

First, it interprets this as a rights-package consisting of the four human rights to the privacy of one's spaces, information about one, one's experiences, and one's decisions. This guides consideration of the moral problems concerning procreational autonomy toward the crucial issue of the privacy of personal decisions and away from irrelevant debates about the privacy of one's spaces, information about one or one's experiences. Second, it defines the core of the human right to the privacy of one's decisions as a moral claim against interference rather than a liberty to make and act on private decisions. By making explicit, the modality of this human right, it centers our thinking about it on what gives it moral force and practical importance in our lives. Third, probably its most original and controversial presupposition is that a moral claim of X against Y consists of a moral duty of Y together with a moral power of X. This makes it clear that the human right to the privacy of one's decisions must be grounded on two very different sorts of moral reasons, duty-imposing reasons and power-conferring reasons. Even if readers remain unconvinced by the reasons suggested in this essay, it should advance our understanding of this right to privacy by challenging them to find more adequate reasons.

"Possessors of the Right to Procreational Autonomy" uses the author's conception of rights to illuminate the limited relevance of moral rights to the question of whether mentally retarded persons and minors ought to possess the legal right to procreational autonomy. Because rights concern the proper allocation of dominion, freedom, and control over some core legal or moral position, only moral agents are possible right holders. Because some, but not all, mentally retarded persons and all very young children lack the capacity to decide and act on the basis of moral reasons, they should not be taken to possess any moral rights. Therefore, other moral reasons are required to decide whether they ought to possess any constitutional right to procreational autonomy.

Thus, there are a variety of ways in which the author's abstract conception of rights might help to answer very specific practical questions about what our medical law ought to be and how individuals are morally permitted or required to decide and act. If this conception is the correct or most adequate way of thinking about legal and moral right, it promises to be useful in reaching the correct or most adequate answers to these questions. If, as is alas possible, the author's general theory of rights is misconceived, it still provides an alternative viewpoint that will be useful by pointing to morally relevant reasons too seldom taken seriously or even considered at all.

The following essays have four primary purposes—to advocate specific moral reforms of existing medical law; to identify some of the most important moral reasons that potential parents, patients, physicians, and hospital administrators should consider when faced with difficult moral choices; to contribute to the mid-level theory of moral rights; and to test the usefulness of the author's general theory of rights in resolving the legal and moral issues posed by particular medical cases and problematic moral decisions. How well they achieve these purposes may best be left to the critical judgment of the reader.

CHAPTER 2

DEFINING THE RIGHTS TO PHYSICIAN-ASSISTED SUICIDE

It is often thought, and with good reason, that jurisprudence is of little or no use to the practicing lawyer or even to the judge. After all, the very general theories proposed by jurists are so abstract that their relevance, if any, to specific legal issues or to particular cases is typically doubtful and at best indeterminate. As a philosopher of law, I remain cautiously hopeful that my speculations do have some practical import. To this end, I have formulated my general theory of rights[1] in terms of Wesley Newcomb Hohfeld's fundamental legal conceptions[2] and their moral analogs. My hypothesis is that this conceptual framework is the most promising way to make explicit the practical relevance of the inarticulate language of rights. I submit this essay to jurists and philosophers as a test of this hypothesis.

During the past few years there have been a number of controversial cases in which physicians and their patients have petitioned the courts to recognize a right to physician-assisted suicide. Unfortunately, these cases have done as much to confuse as to clarify the nature of the right at issue. In the most definitive opinion to date, Chief Justice Rehnquist, speaking for the United States Supreme Court, has complained about the vague and variable language used by and submitted to the United States Court of Appeals, Ninth Circuit, in describing this right.

> Turning to the claim at issue here, the Court of Appeals stated that "[p]roperly analyzed, the first issue to be resolved is whether there is a liberty interest in determining the time and manner of one's death," ... or, in other words "[i]s there a right to die?" .... Similarly, respondents assert a "liberty to choose how to die" and a right to "control of one's final days," ... and describe the asserted liberty as "the right to choose a humane, dignified death" ... and "the liberty to shape death" .... As noted above, we have a tradition of carefully formulating the interest at stake in substantive-due-process cases.[3]

He insisted that only after the right claimed has been defined with some precision could one determine whether it is protected by the Due Process Clause.

He also noted that other courts had described the right to physician-assisted suicide in very different ways.

> See, e.g., *Quill* v. *Vacco*, ... "right to assisted suicide" ...; *Compassion in Dying* v. *Washington* ... "right to suicide," "right to assistance in suicide," and

[1] C. Wellman, *A Theory of Rights* (Totowa, NJ: Rowman & Allanheld, 1985).
[2] W. N. Hohfeld, *Fundamental Legal Conceptions* (New Haven: Yale University Press, 1919).
[3] *Washington* v. *Glucksberg*, 521 U.S. 702 (1997) at 722.

"right to aid in killing oneself"; *People* v. *Kevorkian*... "a right to commit suicide and... whether it includes a right to assistance."[4]

How, then, should the alleged right to physician-assisted suicide be defined? Rehnquist's attempt to provide the careful formulation needed for substantive due process analysis is as follows:

> The Washington statute at issue in this case prohibits "aid[ing] another person to attempt suicide"... and, thus, the question before us is whether the "liberty" specially protected by the Due Process Clause includes a right to commit suicide which itself includes a right to assistance in doing so.[5]

But this formulation raises as many questions as it answers. If the statute at issue prohibits *physicians* from assisting a patient to commit suicide, why is it that the relevant right is the patient's right to commit suicide rather than the physician's right to assist? After all, no Washington statute prohibits any patient from committing suicide. And how could any right to commit suicide include a right to assistance? And why would any right to assistance be limited to assistance from one's physician? Clearly, we need a much more clear and precise definition of the right to physician-assisted suicide before we can identify the issues raised by those who have claimed this right or assess the validity of the legal reasoning in the judicial opinions that have upheld or denied their claims.

I shall attempt to satisfy this need in the pages that follow. I shall, however, limit my subject in two respects. I shall propose a definition of the legal right or rights to physician-assisted suicide but ignore any analogous moral right or rights because doing so would at least double the length of this essay. And I shall attempt to define with precision the right or rights actually at issue in recent court cases, not some ideal right that it might have been wiser for the petitioners to have claimed.

2.1. Constitutional Rights

The right to physician-assisted suicide under United States law might be a constitutional right, a statutory right, or a common law right. Let us begin with those cases in which the plaintiffs are claiming their constitutional rights. If the statute at issue in *Glucksberg* prohibited physicians from assisting anyone to commit suicide, why was the question before the Supreme Court whether the Due Process Clause protects the liberty of patients to commit suicide, including suicide with assistance, *rather than* the liberty of physicians to assist their patients in committing suicide? It was not so limited. The question before the court was whether the Due Process Clause protects both of these liberties. In *Glucksberg*, and in similar cases, not one but two constitutional rights were at issue.

This case originated when two sets of plaintiffs, three physicians and three patients, along with the nonprofit organization Compassion in Dying, sued in the United States District Court seeking a declaration that the State of Washington statute making promoting a suicide attempt a felony is unconstitutional. The physicians declared that they would

[4] *Glucksberg*, note 18 at 723.

[5] *Glucksberg* at 723.

assist their terminally ill suffering patients in ending their lives were it not for Washington's assisted-suicide ban; the patients, who were then in the terminal phases of serious and painful illnesses, declared that they desired assistance in ending their lives.[6] These two sets of plaintiffs sued jointly because the prohibition of assisting a suicide attempt denied physicians the legal liberty of assisting their terminally ill patients to end their suffering by committing suicide and at the same time limited the liberty of those patients to commit suicide with the assistance of their physicians. Hence, the three physicians claimed their constitutional right to assist patients in committing suicide and the three patients asserted their constitutional right to commit suicide with the assistance of their physicians.

These two alleged rights are mutually dependent, although in different ways. A patient cannot have the right to commit suicide with her physician's assistance if her physician has no right to assist because the statute that denies her physician his right to assist would make her an accomplice in his crime were she to engage in physician-assisted suicide. And although some physicians do disobey laws prohibiting assisting another to commit suicide, her opportunity to exercise her right to commit physician-assisted suicide, even if she had it, might well be denied because many physicians are unwilling to disobey the law.

Conversely, the physician's constitutional right to assist a patient to commit suicide will depend upon the patient's constitutional right to commit suicide with her assistance because it will almost certainly be derived from that right. Admittedly, the physician-plaintiffs in *Quill* v. *Vacco* alleged that

> [the] Fourteenth Amendment guarantees the liberty of physicians to practice medicine consistent with their best professional judgment, including using their skills and powers to facilitate the exercise of the decision of competent, terminally ill adults to hasten inevitable death by prescribing suitable medications for the patient to self-administer for that purpose.[7]

But this is a weak argument that is unlikely to convince any court. Nor is there any more plausible independent line of reasoning from constitutional law to the right of physicians to assist their suffering patients to commit suicide. Hence, physicians who wish to claim such a constitutional right must ground their claim upon the patient's constitutional right to physician-assisted suicide, a right for which there are more plausible, although to date unavailing, arguments.

In these respects, as in so many others, a comparison with *Roe* v. *Wade*[8] is illuminating. Although the decision of the Supreme Court in that case is usually said to have established "the right to abortion," in fact it recognized two interdependent constitutional rights—the pregnant woman's right to seek and submit to an abortion and the physician's right to perform an abortion. Moreover, the latter was derived from the former because both were held to be grounded on the pregnant woman's constitutionally protected right to privacy.

First, then, let us formulate a provisional definition of the alleged right of the physician to give assistance to a patient who wishes to commit suicide. Who would be the

---

[6] *Glucksberg* at 707.
[7] *Quill* v. *Vacco*, 80 F.3rd 716 at 719.
[8] 410 U.S. 113 (1973).

right-holders of the constitutional right affirmed in *Compassion in Dying*[9] and denied in *Glucksberg*? Obviously physicians, but probably not all physicians. The physician-respondents were doctors who "occasionally treat terminally ill, suffering patients, and declare that they would assist these patients" were it not for the legal prohibition.[10] And the Court of Appeals had struck down Washington's assisted-suicide ban only "as applied to terminally ill competent adults who wish to hasten their deaths by obtaining medication prescribed by their doctors."[11] Although these formulations are not as explicit as one would wish, they strongly suggest that the right to assist another to commit suicide would be possessed only by attending physicians, those primarily responsible for the care of a patient.

Whom might an attending physician assist? Obviously only one's own patients, patients for whose medical care one has primary responsibility. But by no means all of those.

> In their brief to this Court, the doctors claim not that they ought to have a right generally to hasten patients' imminent deaths, but only to help patients who have made "personal decisions regarding their own bodies, medical care, and, fundamentally, the future course of their lives"... and who have concluded responsibly and with substantial justification that the brief and anguished remainders of their lives have lost virtually all value to them.[12]

In other words, the physician-respondents asserted a right to assist only patients who are mentally competent adults who are terminally ill and are suffering a painful process of dying.

What kinds of assistance would physicians have a right to provide such patients? Certainly not to give the patient a loaded gun with instructions as to where to aim it in order to kill himself most effectively. Presumably not to provide the sorts of devices that Kevorkian has constructed to assist persons who were not his own patients to commit suicide. The most explicit statement of this aspect of the asserted right is as follows:

> Respondents claim that a patient facing imminent death, who anticipates physical suffering and indignity, and is capable of responsible and voluntary choice, should have a right to a physician's assistance in providing counsel and drugs to be administered by the patient to end life promptly .... They accordingly claim that a physician must have the corresponding right to provide such aid ... .[13]

Hence, the physician is probably permitted to assist only by prescribing some lethal medication and advising her patient on how to commit suicide by administering this medication to himself.

In summary, one can provisionally define the asserted physician's constitutional right regarding physician-assisted suicide as the right of an attending physician to prescribe lethal medication and advise an adult mentally competent patient who is terminally ill and suffering a painful dying on how to use this medication to commit suicide.

---

[9] 79 F.3rd 790 (9th Cir. 1996).
[10] *Glucksberg* at 707.
[11] *Glucksberg* at 709.
[12] *Glucksberg* at 754.
[13] *Glucksberg* at 773.

Second, we must now give an illuminating preliminary definition of the alleged patient's constitutional right to physician-assisted suicide. Presumably by far the most authoritative formulations are those of Chief Justice Rehnquist in delivering the opinion of the Supreme Court. After objecting to the descriptions used in previous cases, he wrote:

> ...the question before us is whether the "liberty" specially protected by the Due Process Clause includes a right to commit suicide which itself includes a right to assistance in doing so.[14]

Although this language suggests that the alleged right to physician-assisted suicide is a right that includes a subright, I believe that properly parsed only the words "a right to assistance in doing so" refer to the right we are seeking to define. If so, it would be described as "the right to assistance in committing suicide."

This reading of the text is quickly confirmed by another passage on the very next page.

> The question presented in this case, however, is whether the protections of the Due Process Clause include a right to commit suicide with another's assistance. With this "careful description" of respondents' claim in mind, we turn to *Casey* and *Cruzan*.[15]

This description of the asserted right to physician-assisted suicide may be careful and even accurate, but it is also inarticulate and imprecise. In order to formulate a more helpful definition of this alleged constitutional right, we must turn to the more specific claims of the respondents.

Who would possess the right to physician-assisted suicide? Presumably persons very like the three patient-plaintiffs in the District Court who "were in the terminal phases of serious and painful illnesses" and who declared that they were "mentally competent and desired assistance in ending their lives."[16] Thus, the right-holders of the claimed right, were it recognized, would be mentally competent adults suffering from a painful terminal illness.

What would be the defining content of their right to physician-assisted suicide? The respondents claimed that such patients "should have a right to a physician's assistance in providing counsel and drugs to be administered by the patient to end life promptly."[17] Implicit in this concise description are three distinct components. (1) The qualified patient should have a right to request that her attending physician prescribe lethal medication and advise her on whether, when, and how these drugs should be used, (2) She should have the right to obtain any drugs prescribed, and (3) she should have the right to use these drugs to commit suicide.

Are these three components three aspects of one unified but complex right or the defining cores of three rights that together constitute a rights-package? A rights-package is a set of logically and legally distinct rights relating to a common subject matter. This is surely the case regarding these components. The three kinds of action that define these

[14] *Glucksberg* at 723.
[15] *Glucksberg* at 724.
[16] *Glucksberg*, note 4 at 708.
[17] *Glucksberg* at 773.

components—requesting, obtaining, and using medical assistance—are logically distinct from one another. And these components are legally separate also because a right to one could exist in a legal system without any right to the others, and even were all three real rights, some could be exercised without exercising the others. Thus, a patient could request that her doctor prescribe some lethal medication but not have the prescription filled or, after obtaining the prescribed medication, not use it to commit suicide.

Here, then, we have our preliminary definition of the alleged constitutional right to physician-assisted suicide, at least as it was claimed in *Glucksberg*. It is a rights-package possessed by a mentally competent adult patient suffering from a painful terminal disease consisting of the three rights (1) to request, (2) to obtain, and (3) to use prescribed lethal medication and advice from her attending physician.

2.2. The Modality of the Rights

There are at least four species of legal rights—claim-rights, liberty-rights, power-rights, and immunity-rights.[18] What determines the modality of any right is the modality of its defining core legal position. Thus, the creditor's legal right to be repaid is a claim-right because its core legal position is a legal claim to repayment holding against the debtor. Similarly, the owner's legal right to drive her car whenever she wishes is a liberty-right because at its core is her legal liberty of doing so. In order to give a more complete definition of the alleged physician's right to assist a patient to commit suicide as well as the asserted patient's constitutional right to physician-assisted suicide we must determine their modalities.

First, the physician's alleged right to assist would be a liberty-right. In *Glucksberg* the three physician-claimants were doctors who practiced in Washington. "These doctors occasionally treat terminally ill, suffering patients, and declare that they would assist these patients in ending their lives if not for Washington's assisted-suicide ban."[19] This statute banning anyone from assisting another to commit suicide imposed upon them a legal duty not to assist their patients in ending their lives. What they sought was a declaration by the Supreme Court that the statute is unconstitutional thus eliminating this duty. In other words, in essence they were seeking a legal liberty to assist, that is the absence of a duty not to assist some of their patients to commit suicide.

This interpretation is confirmed by a passage from Justice Souter's concurring opinion.

> In their brief to this Court, the doctors claim not that they ought to have a right generally to hasten patients' imminent deaths, but only to help patients who have made "personal decisions regarding their own bodies, medical care, and, fundamentally, the future course of their lives"... .[20]

The physicians were claiming a right to do something, to help some, but not all, of their patients to hasten their deaths. Because doing so would not be exercising a legal

---

[18] This classification is derived from, but not endorsed by, Wesley Newcomb Hohfeld, supra, p. 71.

[19] *Glucksberg* at 707.

[20] *Glucksberg* at 754.

power, they could not be claiming a power-right. Hence, they must have been claiming a constitutional liberty-right.

More specifically, they were claiming a bilateral liberty-right, a right to assist *or* not assist some of their patients to commit suicide. In *Glucksberg*, no physician-respondent admitted any legal duty to assist any patient to commit suicide; each sought only the liberty of doing so if he so chose. Nor did any patient-respondent allege any such duty; at most a patient assumed that a physician who refused her request for assistance in committing suicide ought to refer her to another physician. It was generally assumed by all parties that physicians ought to be free to follow their own consciences and best medical judgment in such cases.

The physician-respondents grounded their claim upon the Due Process Clause that confers upon all persons an immunity against any State action depriving them of life, liberty, or property without due process of law. Hence, it might be thought that they were claiming an immunity-right to assist their patients. But the purpose of the Due Process Clause is to *protect* the rights to life, liberty, and property. Therefore, what the physicians were claiming was a protected liberty-right, a legal right with a bilateral liberty to assist or not assist as its defining core and an immunity against State legislation denying or unduly burdening this liberty as an associated element.

Second, the patient's asserted constitutional right to physician-assisted suicide is also a liberty-right, more precisely a rights-package consisting of three liberty-rights—the rights (1) to request, (2) to obtain, and (3) to use prescribed lethal medication and advice from one's attending physician. That they were claiming liberty-rights is clear from the language they used in asserting their claim.

> The plaintiffs asserted "the existence of a liberty interest protected by the Fourteenth Amendment which extends to a personal choice by a mentally competent, terminally ill adult to commit physician-assisted suicide."[21]

Moreover, the United States Supreme Court recognized that what was at issue in this case was the existence of a right to some sort of a liberty.

> . . . and, thus, the question before us is whether the "liberty" specially protected by the Due Process Clause includes a right to commit suicide which itself includes a right to assistance in doing so.[22]

And in denying the claim of the respondents, the Court made it clear that they were denying the existence of any constitutional liberty-right.

> That being the case, our decisions lead us to conclude that the asserted "right" to assistance in committing suicide is not a fundamental liberty interest protected by the Due Process Clause.[23]

Hence, the asserted patient's right to physician-assisted suicide would be, were it recognized, a rights-package of liberty-rights.

---

[21] *Glucksberg* at 708.
[22] *Glucksberg* at 723.
[23] *Glucksberg* at 728.

In fact, each of its component rights would be a bilateral liberty. Thus, it would consist of the rights (1) to request or not request, (2) to obtain or not obtain, and (3) to use or not use prescribed lethal medication and advice from one's attending physician.

> They [the patient respondents] seek the option to obtain the services of a physician to give them the benefit of advice and medical help. . . .[24]

> and

> Respondents argued that "the constitutional principle behind recognizing the patient's liberty to direct the withdrawal of artificial life support applies at least as strongly to the choice to hasten impending death by consuming lethal medication". . . .[25]

Because the patient-respondents wanted the liberty to *choose* whether or not to commit physician-assisted suicide, they were asserting a rights-package consisting of three option rights, each defined by a core bilateral liberty.

These liberty-rights would not, of course, necessarily impose any corresponding legal duties upon the attending physician. The logical correlative of a legal liberty is the absence of a contrary legal duty upon the right-holder, not the presence of a legal duty upon any second party.[26] Unfortunately, the right to physician-assisted suicide has sometimes been misinterpreted as a claim-right or set of claim-rights that would impose such a duty or set of duties. For example, Bernard Baumrin has written: "Anyone claiming a right to physician assistance in dying must show that some physician has a duty to satisfy that right."[27] And in a Canadian case in which Sue Rodriguez claimed a fundamental constitutional right to physician-assisted suicide, the Supreme Court of British Columbia denied her claim because it wrongly assumed that recognizing a right to physician-assisted suicide would be "tantamount to imposing a duty upon physicians to assist patients who choose to terminate their own lives. . . ."[28] As far as I am aware, no party who has asserted a constitutional right to physician-assisted suicide before a United States court has alleged or implicitly assumed that it would imply any correlative duty upon the attending physician. Quite the contrary. What all such petitioners have sought has been merely the legal liberty of attending physicians to assist a limited category of patients who wish to commit suicide.

They have, however, assumed that any constitutional right to physician-assisted suicide would be more than a naked liberty or set of liberties. They have argued that it is implied by and thus would be protected by the Due Process Clause. Accordingly, if recognized, the patient's right to physician-assisted suicide would include as associated legal positions immunities against at least (1) any prohibition of committing suicide, (2) any prohibition of requesting, obtaining or using assistance from one's attending physician in doing so, and (3) any prohibition of providing such assistance by that physician.

---

[24] *Glucksberg* at 774.
[25] *Glucksberg* at 725.
[26] Hohfeld, supra, pp. 38–39.
[27] B. Baumrin, Physician, stay thy hand! In M. P. Batten, R. Rhodes, and A. Silvers (eds.), *Physician Assisted Suicide* (New York: Routledge, 1998), pp. 178–179.
[28] *Rodriguez* v. *British Columbia* (*Attorney-General*) 107 D.L.R. (4th) at 351.

2.3. Linkage

In *Glucksberg* and other court cases in which patients asserted their constitutional right to physician-assisted suicide, physicians also asserted a constitutional right to assist terminally ill suffering patients who choose to commit suicide. This is no accident because these two claims are essentially linked. Hence, one test of the correctness of a definition of the alleged patient's right to physician-assisted suicide is how well it can explain the way in which this right is connected with the corresponding right of the physician.

First, the physician's right to assist is, or would be were it recognized, grounded on the patient's right. In claiming their respective rights, the patients appealed directly to the Due Process Clause while the physicians appealed to the Due Process Clause indirectly via the patient's right to physician-assisted suicide.

> Respondents claim that a patient facing imminent death, who anticipates physical suffering and indignity, and is capable of responsible and voluntary choice, should have a right to a physician's assistance in providing counsel and drugs to be administered by the patient to end life promptly .... They accordingly claim that a physician must have the corresponding right to provide such aid ... .[29]

Thus the argument of the physician-respondents was, not that the Washington statute prohibiting assisting another to commit suicide violated their constitutionally protected liberty to practice medicine consistent with their best medical judgment, but that it denied patients their liberty of committing physician-assisted suicide.

Second, this legal reasoning assumed, of course, that the patient's right to physician-assisted suicide somehow requires the existence of the physician's right to provide assistance. How might this be so? The defining cores of the former do not logically imply the defining core of the latter. That is, the patient could have the three legal liberties to request, obtain, and use assistance from her attending physician to commit suicide even though the physician were to have a legal duty not to provide such assistance. This would be true, for example, were the Washington statute prohibiting physicians from assisting their patients to commit suicide not to implicate patients who accepted their assistance as accomplices in any crime and were there no other statute prohibiting patients from requesting, obtaining, or using the assistance from their physician to commit suicide.

Rather, it is because the patient's right to physician-assisted suicide would be a constitutionally protected right that it legally implies the corresponding right of the physician to assist them. Justice Souter in his concurring opinion makes clear the parallel with *Roe* v. *Wade* and other abortion cases in this respect.

> Without physician assistance in abortion, the woman's right would have too often amounted to nothing more than a right to self-mutilation, and without a physician to assist in the suicide of the dying, the patient's right will often be confined to crude methods of causing death, most shocking and painful to the decedent's survivors.[30]

---

[29] *Glucksberg* at 773.
[30] *Glucksberg* at 778.

Thus, although the patient's right to commit suicide, even to commit physician-assisted suicide, could exist without any corresponding right of her physician to assist her, it would be unduly burdened by the absence of the latter right. This is because while a patient who desires to exercise her constitutional right to commit physician-assisted suicide might be fortunate enough to have an attending physician willing to violate the law, many others would be attended by physicians who would refuse assistance out of respect for a statute that prohibits assisting another to commit suicide or out of fear of the penalties they would risk by violating it. Thus, any such statute would, absent some justifying State interest, deny many patients liberty without due process of law. Accordingly, it is the constitutional immunities derived from the Due Process Clause protecting the patient's right to physician-assisted suicide that imply the physician's right to provide assistance. In this way these two rights are linked, not through their defining cores, but through their associated positions, specifically the immunities that protect their core legal liberties.

2.4. Statutory Rights

Not all the recent court cases involving an alleged right to physician-assisted suicide concern an asserted constitutional right. In *Lee* v. *State of Oregon*,[31] a statutory right to physician-assisted suicide was challenged. Let us examine this case, and the Oregon Death With Dignity Act[32] at issue in it, to see how a statutory right to physician-assisted suicide might compare with an analogous constitutional right.

One similarity leaps to mind. The Oregon statute, submitted to the voters as Ballot Measure 16, proposed to create a pair of rights, the patient's right to physician-assisted suicide and the physician's right to assist a patient to do so just as both patients and physicians claimed constitutional rights in *Glucksberg*. On the other hand, the physician's statutory right to assist is not grounded upon the patient's statutory right to physician-assisted suicide as was the case in *Glucksberg*; the two rights were created in the same way by Measure 16 and were defined in large measure by different portions of that statute. Still, there must be some linkage between these two rights, for the kinds of assistance an attending physician has a right to provide must be the same as the kinds of assistance a patient has a right to request, obtain, and use to commit suicide.

Oregon Revised Statute 163.125 establishes the crime of manslaughter in the second degree for either recklessly or intentionally causing or aiding another person to commit suicide. Because this prohibition applies primarily to physicians rather than patients, let us begin by defining the physician's right to assist patients who wish to commit suicide. Oregon Measure 16 did not purport to repeal the Oregon prohibition of assisting suicide; it limited the scope of that statute by creating exceptions to it. Hence, having declared Oregon's Death With Dignity Act unconstitutional, the United States District Court decreed, among other things, that

[31] *Lee* v. *State of Oregon* 891 F. Supp. 1429 (D. Or. 1995).

[32] Oregon Revised Statutes, Section 127.800–127.897 (1997). This statute was submitted to the voters as Ballot Measure 16 in a 1994 referendum, held unconstitutional in *Lee* (judgment later vacated on procedural grounds, then re-approved by the voters in 1997).

> Defendant District Attorney Harcleroad is permanently enjoined from recognizing any exceptions from criminal law created by Oregon Ballot Measure 16 in the exercise of his criminal enforcement duties ....[33]

By creating specific exceptions from the Oregon statute that prohibited assisting another to commit suicide, Measure 16 canceled the duty of attending physicians never to do so and thereby conferred upon them a legal liberty-right to assist some of their patients to commit suicide.

The defining scope of the physician's right to assist is narrowly circumscribed by the Oregon Death With Dignity Act. It includes only the liberty to write a prescription to enable a qualified patient to end his or her life in a humane and dignified manner,[34] presumably the liberty to give the prescription to the patient, and by implication the liberty to be present when the patient takes the prescribed medication.[35] Because Measure 16 does not impose a duty upon the physician to perform any of these actions, each of these defining liberties is a bilateral liberty.

The United States District Court asked, and answered in the negative,

> Is there a rational basis for Oregon to immunize physicians from liability for actions taken in "good faith" under Measure 16, irrespective of any medical community standard which applies to their actions outside Measure 16?[36]

Although this might suggest that the physician's right to assist is an immunity-right, the Court is here questioning, not the defining core liberties of that right, but the associated immunities intended by the Oregon Death With Dignity Act to protect that core liberty or set of liberties. These include immunities from being subject to civil or criminal liability or professional disciplinary action for participating in good faith compliance with this Act,[37] from being subject to censure, discipline, suspension, loss of license, loss of privileges, loss of membership, or other penalty for participating or refusing to participate in good faith compliance with this Act,[38] from being charged with neglect for the provision by an attending physician of medication in good faith compliance with the provisions of this Act,[39] and from being under any legal duty to participate in the provision to a qualified patient of medication to end his or her life in a humane and dignified manner.[40]

Accordingly, the structure of the physician's statutory right to assist created by Oregon's Death With Dignity Act is very similar to the analogous constitutional right claimed by physicians in *Glucksberg*. Each is a right consisting of a set of bilateral liberties to provide or not provide specified kinds of assistance each protected by a number of immunities. However, the precise contents of these core liberties and the protective immunities are defined rather differently.

---

[33] *Lee* at 1439.
[34] Section 3.01(9).
[35] Section 4.01(1).
[36] *Lee* at 1437.
[37] Section 4.01 (1).
[38] Section 4.01(2).
[39] Section 4.01(3).
[40] Section 4.01(4).

The patient's statutory right to physician-assisted suicide created by Measure 16 is a rights-package defined by a set of legal liberties. These include the liberty of a qualified patient to request medication for the purpose of ending his or her life in a humane and dignified manner,[41] to rescind his or her request at any time and in any manner without regard to his or her mental state,[42] to receive a prescription for medication to end his or her life,[43] and by implication to take the prescribed medication to end his or her life in a humane and dignified manner.[44] Because Measure 16 imposes no duty to perform any of these actions, each of these defining core liberties is a bilateral liberty to act or not act in one of these ways.

As one would expect, these defining core liberties are protected by a number of legal immunities. These include immunities against having any provision in a contract, will, or other agreement or any obligation affected by whether a patient may make or rescind a request for medication to end his or her life,[45] against having any sale, procurement, or issuance of any life, health, or accident insurance conditional upon or affected by making or rescinding a request for medication to end one's life or for ingesting such medication,[46] and against having actions in accordance with this act taken as constituting suicide under the law.[47]

Thus, the patient's statutory right to physician-assisted suicide created by Oregon's Death With Dignity Act is very similar to the constitutional right to physician-assisted suicide patients asserted in *Glucksberg*. Although the contents of the legal positions that would have constituted these two rights are defined somewhat differently, each is a rights-package of bilateral legal liberties protected by a set of legal immunities.

2.5. Conclusions

Let us return to Chief Justice Rehnquist's observation that the courts have described the right to physician-assisted suicide in very different ways and that a much more careful formulation is required for any adequate legal analysis. Why is a more precise definition of this right needed? As Rehnquist pointed out, one cannot identify the precise legal issue before the court until one knows exactly what right is being asserted by the plaintiffs or the respondents claiming a right to physician-assisted suicide. Presumably the relevance or irrelevance of the arguments presented by the parties before the court will depend upon the issue at hand.

Again, the adequacy of the grounds for any claimed right to physician-assisted suicide, either constitutional or statutory, will necessarily depend upon exactly how they are connected with the precise content of the right being claimed. Similarly, the legal implications of any right to physician-assisted suicide, were it to be upheld, would depend

[41] Sections 2.01 and 3.06.
[42] Section 3.07.
[43] Section 3.04.
[44] Section 4.01(1).
[45] Section 3.12.
[46] Section 3.13.
[47] Section 3.14.

upon precisely how this right is formulated by the claimants and then by the court. It is these legal implications, such as the kinds of legal assistance physicians might be at liberty to provide and under what circumstances, that matter in medical practice to both patients and physicians; and it is potentially undesirable implications that might deter a court from recognizing any such right.

Finally, whether the legal reasoning of the Supreme Court in *Glucksberg* is valid, whether future plaintiffs might win their case and whether future attempts to legislate a statutory right to physician-assisted suicide could withstand constitutional scrutiny will also depend on precisely how this right is defined.

What have we discovered that is relevant to these important concerns in the process of defining the different rights to physician-assisted suicide claimed in *Glucksberg* v. *Washington* and *Lee* v. *Oregon*? First, it is a serious mistake to think only of *the* legal right to physician-assisted suicide as though one and the same right were at issue in every case. Those who asserted a constitutional right to physician-assisted suicide formulated their legal claims differently from case to case, and no doubt different state legislatures would enact somewhat different versions of a statutory right to physician-assisted from state to state. In any event, the right defined by Oregon's Death With Dignity Act is defined more narrowly and in greater detail than the constitutional right asserted by the respondents in *Glucksberg*.

Second, in spite of these differences, all the patients' rights to physician-assisted suicide at issue in recent court cases have an essentially similar structure. All are rights-packages defined by a set of core bilateral legal liberties protected by associated legal immunities of various kinds. It is a prejudicial misinterpretation to imagine that any of these rights is a claim-right that would impose upon the attending physician or upon the state a correlative legal duty to assist a patient who wishes to commit suicide. At most, some rights to physician-assisted suicide might contain an associated duty of the physician who refuses to assist a patient who wishes to exercise her right to physician-assisted suicide to refer the patient to another physician.

Third, any right to physician-assisted suicide will secure what some terminally ill suffering patients seek only if it is paired with a corresponding right of the physician to provide the assistance the patient needs to enable her to commit suicide. Although any constitutional right of the physician to assist will be tied to the patient's right to commit physician-assisted suicide more intimately than is the case with the analogous pair of statutory rights, the former must at least permit the physician to provide the kinds of assistance that the latter permits the patient to request, obtain, and use.

Fourth, the potential grounds of any constitutional right to physician-assisted suicide will differ markedly from and be much more controversial than the grounds of any comparable statutory right. The latter will be grounded directly upon some enacted legislation; constitutional law will be relevant only insofar as it might challenge the validity of the state statute. Although the most plausible grounds for any constitutional right to physician-assisted suicide are the constitutional right to privacy or the common law right to bodily integrity, no arguments from these grounds, or from the right to equal protection of the laws, has yet convinced the United States Supreme Court. In the future, those who claim any such right would do well to take seriously Chief Justice Rehnquist's demand for a much more careful formulation of the asserted right in terms that show clearly how it can be derived from firm constitutional grounds and why it is not over-ridden by the serious state interests to which the courts appeal to deny such claims.

My primary purpose in this essay has been to show by example how one might best go about defining a right to physician-assisted suicide in order to clarify the legal issues it would raise and help to decide these issues in a reasonable and responsible manner. I do not insist that my definitions of the rights to physician-assisted suicide claimed in recent court cases are correct in every detail. On matters of substantive law, one needs the opinion of a lawyer, not a philosopher of law. What I have tried to show is that definitions of the kind I have formulated would be useful to trial lawyers and would improve the judicial reasoning of the courts.

CHAPTER 3

GLUCKSBERG v. COMPASSION

Because Judge Reinhardt's arguments in *Compassion in Dying v. State of Washington*[1] present by far the most plausible case for a constitutional right to physician-assisted suicide to date, one would like to know whether the reasoning in *Washington v. Glucksberg*[2] that reversed the decision of the Ninth Circuit court adequately rebuts those arguments. This question remains important, not only because new plaintiffs will probably appeal to variations on Reinhardt's arguments in claiming this right, but also because the Supreme Court itself has recognized that the debate about the morality, legality, and practicality of physician-assisted suicide will and should continue. The purpose of this essay will be to answer this question. To simplify the structure of my exposition and evaluation, I will focus on the reasoning of Chief Justice Rehnquist, who delivered the opinion of the Court in *Glucksberg*, and will introduce other materials only when and as they become relevant.

3.1. History, Tradition, and Practice

After tracing the origins of this case, Rehnquist examined our nation's history, legal traditions, and practices.[3] He reported that for over 700 years the Anglo-American common-law tradition has punished or at least disapproved of both suicide and assisted suicide. Suicide was a crime under English common-law and the American colonies mostly adopted the common-law approach. The movement away from the harsh penalties for suicide represented a growing belief that it is unfair to punish the suicide's family for his wrongdoing, not any acceptance of the act of suicide. The earliest American statute explicitly prohibiting assisting suicide was enacted in 1828, and today most states have such statutes in force. These bans have recently been re-examined and generally reaffirmed. He concluded:

> Attitudes toward suicide itself have changed since Bracton, but our laws have consistently condemned, and continue to prohibit, assisting suicide.[4]

The relevance of this historical background to Rehnquist's legal reasoning remains to be seen.

Comparing this historical examination with the comparable portions of Reinhardt's opinion,[5] one notices two differences. While Rehnquist emphasized specifically legal

[1] 79 F. 3rd 790 (9th Cir. 1996).
[2] 521 U.S. 702 (1997).
[3] *Glucksberg* at 710–719.
[4] *Glucksberg* at 719.
[5] *Compassion* at 806–812.

history, Reinhardt provided a much broader survey of philosophical, literary, and cultural attitudes toward suicide and assisted suicide. And Reinhardt added a discussion, missing in Rehnquist, of current societal attitudes regarding terminally ill patients and accepted ways of avoiding painful, undignified, and inhumane deaths. Whether these differences undermine the reasoning in *Glucksberg* depends upon how Rehnquist used historical information in justifying his conclusion.

3.2. Due Process Analysis

Chief Justice Rehnquist began his legal reasoning by granting an essential premise of the patients claiming a right to physician-assisted suicide. The Due Process Clause, "nor shall any State deprive any person of life, liberty, or property, without due process of law," guarantees more than fair process and the absence of physical restraint. It also guarantees that no person shall be deprived of life, liberty, or property by arbitrary or unreasonable legislation. But he added that the Supreme Court has always been reluctant to expand the scope of substantive due process because of the danger of substituting the subjective policy preferences of its members for the objectively defined liberty protected by the Constitution.

He cited a long line of cases to show that the Due Process Clause provides heightened protection against any governmental interference with certain fundamental rights including the rights to marry, to have children, to direct the upbringing of one's children, to bodily integrity, to abortion, and to refuse unwanted lifesaving medical treatment.[6] He found in these cases the proper method of identifying those fundamental rights not specified in the Bill of Rights.

> Our established method of substantive-due-process analysis has two primary features: First, we have regularly observed that the Due Process Clause specially protects those fundamental rights and liberties which are, objectively, "deeply rooted in this Nation's history and tradition". . . . Second, we have required in substantive-due-process cases a "careful description" of the asserted fundamental liberty interest.[7]

The advantages of this method are that it tends to rein in the subjective elements that are present in due-process judicial review and to avoid the need for a complex balancing of competing interests in every case.

Does this imply that only those liberty interests that can be found in our Nation's history and tradition *as carefully described* may be recognized as fundamental rights? If so, this proposed method of analysis is very different from the legal standard defended by Reinhardt.

> Thus, while historical analysis plays a useful role in any attempt to determine whether a claimed right or liberty interest exists, earlier legislative or judicial recognition of the right or interest is not a *sine qua non*. In *Casey*, the Court made

[6] *Glucksberg* at 720.
[7] *Glucksberg* at 720–721.

> it clear that the fact that we have previously failed to acknowledge the existence of a particular liberty interest or even that we have previously prohibited its exercise is no barrier to recognizing its existence.[8]

Reinhardt pointed out that were history the sole standard, the Virginia antimiscegenation statute that the Supreme Court unanimously overturned in *Loving v. Virginia*[9] would still be in force. Much the same could be said of the Texas antiabortion statute that the Supreme Court declared unconstitutional in *Roe v. Wade*.[10]

Has Rehnquist misinterpreted the established method of substantive-due-process analysis? In my judgment, this depends upon the precise relationship between the two primary features upon which he insists—that the fundamental right be deeply rooted in this Nation's history and that the liberty interest be carefully described. As Reinhardt correctly argued, the "particular liberty interest," the specific right claimed by the petitioners, need not be found in the legal history of our Nation. At the same time, one must be able to find something, either a recognized legal principle or previous decisions in analogous cases, in our history and tradition to provide an objective ground for the claimed right. Thus, much will depend upon where and how history is introduced into the legal reasoning of the court, whether one is identifying the fundamental right claimed by the petitioners or identifying the legal sources to which one appeals to ground the asserted right.

3.3. Defining the Issues

In *Compassion*, three terminally ill patients, four physicians, and a nonprofit organization brought suit against the state of Washington seeking a declaration that its statute prohibiting causing or aiding another person to commit suicide is unconstitutional because it violates a liberty interest protected by the Due Process Clause. Hence, Judge Reinhardt recognized that there were two issues before the Ninth Circuit Court of Appeals, whether a patient who is terminally ill has a constitutionally protected liberty interest in hastening her death and, if so, whether the state of Washington may restrict its exercise by banning physician-assisted suicide.[11]

Chief Justice Rehnquist rejected the way in which Reinhardt had posed the question at issue as inconsistent with the established method of substantive-due-process analysis. What was required was a more careful description of the asserted liberty interest.

> The question presented in this case, however, is whether the protections of the Due Process Clause include a right to commit suicide with another's assistance. With this "careful description" of respondents' claim in mind, we turn to *Casey* and *Cruzan*.[12]

Does this formulation prejudice the issue against those claiming a right to physician-assisted suicide?

---

[8] *Compassion* at 805.
[9] 388 U.S. 1 (1967).
[10] 410 U.S. 113 (1973).
[11] *Compassion* at 793.
[12] *Glucksberg* at 724.

The following passage suggests that Reinhardt might well be interpreted to argue that it does.

> Properly analyzed, the first issue to be resolved is whether there is a liberty interest in determining the time and manner of one's death. We do not ask simply whether there is a liberty interest in receiving "aid in killing oneself" because such a narrow interest could not exist in the absence of a broader and more important underlying interest—the right to die.[13]

Rehnquist's reformulation of the question before the Supreme Court might be thought to be objectionable, not because it was a careful description of the asserted liberty interest, but because it defined this interest too narrowly.

I do not believe that one needs to object to the way in which Rehnquist has posed this issue. The respondents had asserted a constitutionally protected right to physician-assisted suicide that could be described, although not very carefully, as "a right to commit suicide with another's assistance." Moreover, the issue posed was whether the protections of the Due Process Clause "include" this right, which leaves open the question of whether this narrowly defined right can be derived from a more broadly defined liberty interest. Whether Reinhardt has correctly described any such broader underlying liberty interest also remains open. Therefore, in assessing the reasoning of *Glucksberg*, one must look both to the accuracy of its definition of the asserted right to physician-assisted suicide and to its description of the grounds to which the respondents appealed in asserting this right.

3.4. No Fundamental Right

First, having defined the issue before the Supreme Court, Rehnquist brought to bear his examination of our Nation's history, tradition, and practice.

> We now inquire whether this asserted right has any place in our Nation's traditions. Here, as discussed above, . . . we are confronted with a consistent and almost universal tradition that has long rejected the asserted right, and continues explicitly to reject it today, even for terminally ill, mentally competent adults.[14]

Granted that legal tradition has long rejected the asserted right to physician-assisted suicide, has that tradition accepted some underlying liberty interest upon which a claimant could ground that asserted right?

Rehnquist ignored this possibility because he wrongly assumed that the arguments in *Compassion* do not rely upon history in any serious manner.

> Respondents contend, however, that the liberty interest they assert is consistent with this Court's substantive-due-process line of cases, if not with this nation's history and practice.[15]

---

[13] *Compassion* at 801.
[14] *Glucksberg* at 723.
[15] *Glucksberg* at 723–724.

However, their contention was not merely that the liberty interest they assert is consistent with, but that it is supported by and grounded in general upon a line of cases, a line long enough to itself constitutes a significant part of the legal history of our constitutional law.

It is clear from the place in their respective opinions each cited the leading due process cases that Rehnquist and Reinhardt used history very differently. Rehnquist cited "a long line of cases"[16] primarily to justify his description of "our established method of substantive-due-process analysis" regarding fundamental rights.[17] However, Reinhardt cited essentially the same line of cases as the first of three authoritative legal sources for the protected liberty interest upon which the constitutional right to physician-assisted suicide is grounded.

> A common thread running through these cases is that they involve decisions that are highly personal and intimate, as well as of great importance to the individual.... Accordingly, we believe the cases from *Pierce* through *Roe* provide strong general support for our conclusion that a liberty interest in controlling the time and manner of one's death is protected by the Due Process Clause of the Fourteenth Amendment.[18]

Because Rehnquist virtually ignored this part of Reinhardt's reasoning, his opinion cannot be said to have refuted it.

Second, instead he moved directly to the other two parts of the reasoning in *Compassion*.

> The question presented in this case, however, is whether the protections of the Due Process Clause include a right to commit suicide with another's assistance. With this "careful description" of respondent's claim in mind, we turn to *Casey* and *Cruzan*.[19]

Even here he paid no attention to the way in which Reinhardt had used this line of "prior court decisions" to justify his interpretation of these "two relatively recent decisions" of the Supreme Court.[20]

Reinhardt's argument from Casey[21] rests squarely upon three sentences in that opinion.

> These matters, involving the most intimate and personal choices a person may make in a lifetime, choices central to personal dignity and autonomy, are central to the liberty protected by the Fourteenth Amendment. At the heart of liberty is the right to define one's own concept of existence, of meaning, of the universe, and of the mystery of human life. Beliefs about these matters could not define the attributes of personhood were they formed under compulsion of the State.[22]

---

[16] *Glucksberg* at 720.
[17] *Glucksberg* at 720.
[18] *Compassion* at 813.
[19] *Glucksberg* at 724.
[20] *Compassion* at 812 and 813.
[21] *Planned Parenthood v. Casey*, 505 U.S. 833.
[22] *Casey* at 698, quoted in *Compassion* at 813.

Reinhardt introduced this quotation by pointing out that it was not an *ad hoc* formulation but was based upon the Supreme Court's survey of its prior decisions.

Reinhardt then applied this formulation to the situation of many terminally ill patients.

> Like the decision of whether or not to have an abortion, the decision how and when to die is one of "the most intimate and personal choices a person may make in a lifetime," a choice "central to personal dignity and autonomy." A competent terminally ill adult, having lived nearly the full measure of his life, has a strong liberty interest in choosing a dignified and humane death rather than being reduced at the end of his existence to a childlike state of helplessness, diapered, sedated, incontinent.[23]

From this liberty interest he derived the patient's right to physician-assisted suicide, just as from a woman's liberty interest in choosing whether to bear or beget a child the Supreme Court in *Roe* had derived the constitutionally protected right to abortion.

The crux of Rehnquist's reply in *Glucksberg* is contained in one footnoted sentence.

> By choosing this language, the Court's opinion in *Casey* described, in a general way and in light of our prior cases, those personal activities and decisions that this Court has identified as so deeply rooted in our history and traditions, or so fundamental to our concept of constitutionally ordered liberty, that they are protected by the Fourteenth Amendment.[24]

Note 19 purported to show that the liberties specified in the cited cases were recognized as fundamental rights simply because they were seen to be deeply rooted in our history and traditions, not because they were instances of the general formulation quoted by Reinhardt. To my mind, this examination of the relevant cases is too selective and too incomplete to be convincing. Moreover, if the language of *Casey* did not formulate a general principle implicit in those cases, why does one find similar language time and again in those opinions of the Supreme Court? This possibility cannot be rejected out of hand.

Third, he then dismissed the essential premise of Reinhardt's argument from *Casey* in these words:

> That many of the rights and liberties protected by the Due Process Clause sound in personal autonomy does not warrant the sweeping conclusion that any and all important, intimate, and personal decisions are so protected . . . and *Casey* did not suggest otherwise.[25]

But *Casey* did suggest, even if it did not conclusively prove, otherwise. This is precisely why Chief Justice Rehnquist, joined by Justices White, Scalia, and Thomas dissented in large part from the opinion of the Court in that case. Hence, Rehnquist's brief and inadequate reply in *Glucksberg* to Reinhardt's argument must be read in the light of his opinion in *Casey*.

---

[23] *Compassion* at 813–814.

[24] *Glucksberg* at 727.

[25] *Glucksberg* at 727.

In the relevant portion of that opinion, Rehnquist is arguing that *Roe v. Wade*[26] was wrongly decided and should be over-ruled.

> In Roe v. Wade, the Court recognized a "guarantee of personal privacy" which "is broad enough to encompass a woman's decision whether or not to terminate her pregnancy."... We are now of the view that, in terming this right fundamental, the Court in Roe read the earlier opinions upon which it based its decision much too broadly. Unlike marriage, procreation and contraception, abortion "involves the purposeful termination of potential life."... The abortion decision must therefore "be recognized as sui generis, different in kind from the others that the Court has protected under the rubric of personal or family privacy and autonomy."[27]

In this passage, Rehnquist argued that even if the earlier substantive-due-process decisions of the Supreme Court had recognized a fundamental right to privacy, which they had not, this right would not be broad enough to ground a constitutional right to abortion because the abortion decision is not a purely private decision; it is the decision to kill another person, or at least a potential person.

I find this argument almost persuasive. In the line of privacy cases from *Griswold* to *Roe*, the crucial sentence is probably one in *Eisenstadt v. Baird.*[28]

> If the right of privacy means anything, it is the right of the individual, married or single, to be free from unwarranted governmental intrusion into matters so fundamentally affecting a person as the decision whether to bear or beget a child.[29]

But this definition of a private decision is inadequate, and for the reason that Rehnquist gave. It is not sufficient that the decision fundamentally affect the life of one person for better or worse; it must also *not* affect the life of another person equally fundamentally. Otherwise, a husband's decision to save the expense of a divorce by shooting his wife would be a private decision. For similar reasons, it would be a mistake to define a private decision simply as one that is intimate, personal, and important in the way that *Casey* and *Compassion* often suggest.

Nevertheless, this argument is not a decisive reply to Reinhardt, for whether a terminally ill patient dies now rather than after a few additional months of suffering will typically affect the life of the patient very much more profoundly than it will affect the lives of any others. Therefore, it remains possible that a right to physician-assisted suicide can be grounded on some fundamental liberty interest, such as the patient's interest in privacy, protected by the Due Process Clause.

Fourth, Rehnquist advanced another and more troubling argument against this possibility.

---

[26] 410 U.S. 113.
[27] *Casey* at 764.
[28] 405 U.S. 438 (1972).
[29] *Eisenstadt* at 453.

> We have held that a liberty interest protected under the Due Process Clause of the Fourteenth Amendment will be deemed fundamental if it is "implicit in the concept of ordered liberty."... Three years earlier... we referred to a "principle of justice so rooted in the traditions and conscience of our people as to be ranked as fundamental."... These expressions are admittedly not precise, but our decisions implementing this notion of "fundamental" rights do not afford any more elaborate basis on which to base such a classification.[30]

In other words, a careful examination of the long line of due process cases shows that the only established criteria for the recognition of a liberty interest as fundamental are either that it is implicit in the concept of ordered liberty or that it is deeply rooted in the traditions and conscience of our people. They do not reveal any general principle, such as the ones proposed in *Eisenstadt* or in the opinion of the Court in *Casey*, that is consistently accepted in all or even most of these various cases.

But is this true? Reinhardt had examined this line of due process cases and concluded that "a common thread running through these cases is that they involve decisions that are highly personal and intimate, as well as of great importance to the individual."[31] And Justice Blackmun had previously rejected Rehnquist's reading of these cases in his opinion in *Casey*.

> The Chief Justice's criticism of Roe follows from his stunted conception of individual liberty. While recognizing that the Due Process Clause protects more than simple physical liberty, he then goes on to construe this Court's personal-liberty cases as establishing only a laundry list of particular rights, rather than a principled account of how these particular rights are grounded in a more general right of privacy.[32]

Well, what does a careful examination of the long line of due process cases really show? Because Rehnquist gave only a superficial survey of these cases, he has not refuted the argument that Reinhardt advanced in *Compassion*. On the other hand, Reinhardt provided even less of an examination of the relevant cases to support his formulation of the general principle he alleged to be implicit in them. The hard work of finding some principle of constitutional law in all or most of the long line of due process cases remains to be done.

Those who are tempted to accept the general formulation advanced in the opinion of the court in *Casey* and relied upon by Reinhardt can point out that it echoes the language of many of these cases. The question remains as to precisely how this language was used in the legal reasoning of the Supreme Court in those cases. Is some such general principle essential to the justification of the holdings in due process cases or is it merely incidental to arguments that appeal entirely to the criteria of the concepts of ordered liberty and of liberties deeply rooted in our traditions and legal history? Rehnquist justifies his selection of the criteria of fundamental rights by quoting some of the language of earlier decisions of the Supreme Court. But those who find some general principle in

[30] *Casey* at 763.
[31] *Compassion* at 813.
[32] Casey at 756.

those decisions also quote from them. A few citations taken out of context prove nothing. What is required, and what neither Rehnquist nor Reinhardt provided, is a detailed examination of exactly how the quoted sentences or phrases have been used in the reasoning of the Court. Therefore, although Rehnquist has not adequately rebutted Reinhardt's appeal to Casey, neither has Reinhardt developed his argument fully enough to render it persuasive.

Fifth, the other relatively recent decision to which Justice Reinhardt appealed was *Cruzan v. Director, Missouri Department of Health.*[33]

> Writing for the majority that included Justices O'Connor and Scalia, Chief Justice Rehnquist said that those [cited] cases helped answer the first critical question at issue in *Cruzan*, stating: "The principle that a competent person has a constitutionally protected liberty interest in refusing unwanted medical treatment may be *inferred* from our prior decisions."[34]

and

> In her concurrence, Justice O'Connor explained that the majority opinion held (implicitly or otherwise) that a liberty interest in refusing medical treatment extends to all types of medical treatment from dialysis or artificial respirators to the provision of food and water by tube or other artificial means.[35]

From these two premises, Reinhardt drew his conclusion that *Cruzan*, by recognizing a liberty interest that includes the refusal of artificial provision of life-sustaining food and water, necessarily recognizes a liberty interest in hastening one's own death.[36] But precisely how is this conclusion supposed to establish the asserted right to physician-assisted suicide?

One possible interpretation, the most plausible one, is suggested by the next two sentences.

> *Casey* and *Cruzan* provide persuasive evidence that the constitution encompasses a due process liberty interest in controlling the time and manner of one's death—that there is, in short, a constitutionally recognized "right to die." Our conclusion is strongly influenced by, but not limited to, the plight of mentally competent, terminally ill adults.[37]

Perhaps, then, Reinhardt is arguing that *Cruzan* recognized a constitutionally protected right to die and that the right to die implies the right to physician-assisted suicide, at least for competent terminally ill adults.

Rehnquist did not deny that the Supreme Court in fact recognized a right to die in Cruzan, although he seems to have pointed out that it did not use that terminology in its reasoning.

---

[33] 497 U.S. 261

[34] *Cruzan* at 278, quoted in *Compassion* at 814 (emphasis added by Reinhardt).

[35] *Compassion* at 815.

[36] *Compassion* at 816.

[37] *Compassion* at 816.

> This is the first case in which we have been squarely presented with the issue whether the United States Constitution grants what is in common parlance referred to as a "right to die."[38]

Notice that the expression "right to die" is enclosed in quotation marks and said to belong to common parlance not the language of constitutional law.

Instead, Rehnquist rejected Reinhardt's definition of the right to die as "a due process liberty interest in controlling the time and manner of one's death." This very general conception reflects the mistaken assumption that all intimate, personal, and important decisions are protected by the Due Process Clause.

> The right assumed in *Cruzan*, however, was not simply deduced from abstract concepts of personal autonomy. Given the common-law rule that forced medication was a battery, and the long legal tradition protecting the decision to refuse unwanted medical treatment, our assumption was entirely consistent with this Nation's history and constitutional traditions.[39]

On my reading of *Cruzan*, this is entirely correct. One may hold that in this case the Supreme Court did recognize the right to die, but one must not inflate this right beyond the right to refuse life-prolonging medical treatment, one aspect of the right *not* to be treated without one's consent. It is hard to see how such a limited negative right to die could imply any right to physician-assisted suicide strong enough to ground the legal liberty of a physician to provide medication to enable her patient to kill himself.

There is, however, another possible interpretation of Reinhardt's argument from *Cruzan*, one suggested by the respondents' brief.

> Respondents contend that in *Crusan* we "acknowledged that competent, dying persons have the right to direct the removal of life-sustaining medical treatment and thus hasten death," Brief for Respondents 23, and that "the constitutional principle behind recognizing the patient's liberty to direct the withdrawal of artificial life support applies at least as strongly to the choice to hasten impending death by consuming lethal medication," *id.* at 26.[40]

Here the right to physician-assisted suicide is said to be grounded, not on the right to die itself, but on the constitutional principle behind that right.

Well, what is that underlying constitutional principle? According to Rehnquist, it is the common-law doctrine of informed consent.

> We began with the observation that "[a]t common law, even the touching of one person by another without consent and without legal justification was a battery."... We then discussed the related rule that "informed consent is generally required for medical treatment."... After reviewing a long line of relevant cases, we concluded that "the common-law doctrine of informed consent is viewed as generally encompassing the right of a competent individual to refuse medical treatment."... Next, we reviewed our own cases on the subject, and stated that

---

[38] *Cruzan* at 497.
[39] *Glucksberg* at 725.
[40] *Glucksberg* at 725.

"[t]he principle that a competent person has a constitutionally protected liberty interest in refusing unwanted medical treatment may be inferred from our prior decisions."[41]

As far as I can see, this is a perfectly accurate report of the reasoning of the Supreme Court in *Cruzan*. And, as I have just suggested, the right not to be treated without one's consent is too limited and negative to ground any right to physician-assisted suicide. Therefore, Rehnquist has fully rebutted Reinhardt's argument from *Cruzan* on either or both plausible interpretations of it.

Sixth, but Reinhardt argued from both *Cruzan* and *Casey*, and combining his two arguments might well enable him to challenge Rehnquist's insistence that the constitutional principle behind the right to die is purely and simply the common-law doctrine of informed consent. *Casey* had relied upon the line of cases recognizing a constitutionally protected right to privacy. And in *Cruzan*, Rehnquist had reported that in the *Quinlan*[42] case "the New Jersey Supreme Court granted the relief, holding that Karen had a right of privacy grounded in the Federal Constitution to terminate treatment." He also recognized that in *Saikewicz*,[43] "the Supreme Judicial Court of Massachusetts relied on both the right of privacy and the right of informed consent to permit the withholding of chemotherapy. . . ."[44] Although the Supreme Court in *Cruzan* did not rely on the right to privacy, Reinhardt could argue that the principle implicit in the line of privacy cases is the constitutional principle behind the right to die and that this principle grounds the right to physician-assisted suicide.

It is true that *Cruzan* followed the example of *In re Storar*[45] and *In re Conroy*[46] by grounding the right to refuse medical treatment on the common-law right to informed consent rather than the constitutional right to privacy. But because in all three decisions the respective courts were granting, not rejecting, the asserted right to refuse treatment, they did not need to decide whether it was also grounded on the right to privacy. But here in *Glucksberg* Rehnquist cannot ignore this possibility because he is writing for a Supreme Court denying the fundamental liberty interest asserted by the respondents. In this respect, he has not rebutted the argument of Reinhardt if this is interpreted to appeal to the constitutional principle of privacy behind the right to die. It remains possible, although not proven, that this principle does ground a fundamental right to physician-assisted suicide.

3.5. Statute Not Arbitrary

Rehnquist cannot rest his case here, for even when no fundamental liberty-right is at stake, the Due Process Clause requires that any liberty-limiting statute must not be arbitrary.

---

[41] *Glucksberg* at 724–725.
[42] *In the Matter of Quinlan*, 70 N.J. 10.
[43] *Superintendent of Belchertown State School v. Saikewicz*, 373 Mass. 728.
[44] *Cruzan* at 271.
[45] 52 N.Y.2nd 363.
[46] 98 NJ 321.

> The History of the law's treatment of assisted suicide in this country has been and continues to be one of the rejection of nearly all efforts to permit it. That being the case, our decisions lead us to conclude that the asserted "right" to assistance in committing suicide is not a fundamental liberty interest protected by the Due Process Clause. The Constitution also requires, however, that Washington's assisted suicide ban be rationally related to legitimate government interests.[47]

Rehnquist then asserted that the Washington statute promotes the interests in the preservation of life, in preventing suicide, in protecting the integrity and ethics of the medical profession, in protecting vulnerable groups, and in not starting down the path to voluntary and even involuntary euthanasia. He concluded:

> We need not weigh exactly the relative strengths of these various interests. They are unquestionably important and legitimate, and Washington's ban on assisted suicide is at least reasonably related to their promotion and protection.[48]

And here Rehnquist rested his case.

But is there really no need to weigh the strengths of these state interests? Rehnquist's reasoning in the opinion of the Court consists of a two-stage argument assuming that the Due Process Clause protects only fundamental liberty-rights or nonfundamental liberties that may be limited whenever doing so might reasonably be thought by the legislature to promote some conceivable legitimate state interest. Reinhardt had challenged this assumption.

> Recent cases, including *Cruzan*, suggest that the Court may be heading towards the formal adoption of the continuum approach, along with a balancing test, in substantive due process cases generally. If so, there would no longer be a two-tier or three-tier set of tests that depends on the classification of the right or interest as fundamental, important, or marginal. Instead, the more important the individual's right or interest, the more persuasive the justifications for infringement would have to be.[49]

I doubt, however, that Reinhardt's challenge need be taken very seriously. First, it is not at all clear that the Supreme Court is prepared to reject its traditional two-tiered or more recent three-tiered approach to due process cases. Second, and more crucial, one could adopt a continuum approach and easily argue that the legitimate state interests promoted by a statute prohibiting assisted suicide are important enough to outweigh the individual's liberty interest in physician-assisted suicide. This is illustrated by the opinion of Justice Souter, who recognized most fully the strength of the reasoning in *Compassion* but concurred in the opinion of the Court.

> In my judgment, the importance of the individual interest here, as within that class of "certain interests" demanding careful scrutiny of the State's contrary claim... cannot be gainsaid. Whether that interest might in some circumstances,

---

[47] *Glucksberg* at 728.
[48] *Glucksberg* at 735.
[49] *Compassion* at 804.

> or at some time, be seen as "fundamental" to the degree entitled to prevail is not, however, a conclusion that I need to draw here, for I am satisfied that the State's interests described in the following section are sufficiently serious to defeat the claim that its law is arbitrary or purposeless.[50]

In short, even if Rehnquist should have weighed the relative strengths of the several governmental interests promoted by the Washington statute, he could easily have done so in a manner that would have supported the holding of the Court in this case.

3.6. The Holding of the Court

In concluding, Chief Justice Rehnquist asserted the holding of the United States Supreme Court in this case.

> We therefore hold that Wash. Rev. Code § 9A.36.060(1) (1994) does not violate the Fourteenth Amendment, either on its face or "as applied to competent terminally ill adults who wish to hasten their deaths by obtaining medication prescribed by their doctors."[51]

Notice that this holding goes beyond Rehnquist's reasoning in this case because it ignores the Equal Protection Clause. But presumably it presupposes his reasoning in *Vacco v. Quill*[52] decided at the same time. Accordingly, Rehnquist announced the action of the Supreme Court in reversing the decision of the Ninth Circuit Court of Appeals in *Compassion*.

What is the intended import of this holding? Presumably to rule out any constitutional right to physician-assisted suicide. But this seems to go beyond the combined reasoning in both of these cases, for there are other provisions of the Constitution to which a petitioner could appeal in claiming such a right. A safer and more conservative conclusion would have been simply that the arguments presented to the Supreme Court in this case fail to establish any such right. In any event, the Supreme Court did not intend to hold that a legal right to physician-assisted suicide would necessarily violate the United States Constitution.

> Throughout the Nation, Americans are engaged in an earnest and profound debate about the morality, <u>legality</u>, and practicality of physician-assisted suicide. Our holding permits this debate to continue, as it should in a democratic society.[53]

The concurring opinion of Justice Souter indicates that the option of a statutory right to physician-assisted suicide remains open.[54]

---

[50] *Glucksberg* at 782.
[51] *Glucksberg* at 735.
[52] 521 U.S. 793 (1997).
[53] *Glucksberg* at 735, my emphasis.
[54] *Glucksberg* at 788–789.

3.7. My Conclusion

But is this intended conclusion correct? Only if Rehnquist's reasoning in *Glucksberg* fully rebuts the arguments of Reinhardt in *Compassion*. For reasons that I have explained, I believe that it does not. (1) In his review of our nation's legal history and practice, Rehnquist ignored Reinhardt's claim that a long line of Supreme Court cases provide support for the conclusion that a liberty interest in controlling the time and manner of one's death is protected by the Due Process Clause. (2) Rehnquist has not shown that the liberties specified in the cases he did examine were selected simply because they are deeply rooted in our history and traditions and not at all because they are instances of the general principle formulated by Reinhardt. (3) Although Rehnquist was correct in asserting that not every decision that fundamentally affects the life of an individual for better or worse is protected by the constitutional right to privacy, his objection to the reasoning of the Supreme Court in *Roe v. Wade* is not a decisive reply to Reinhardt. (4) Rehnquist's examination of the long line of privacy cases was too superficial to show that they have established only a list of particular rights rather than a principled account of how these rights are grounded on a more general right to privacy. (5) Although he was correct in asserting that the constitutional right to die can be and in *Cruzan* was grounded on the common-law right to informed consent, Rehnquist was not justified in this case in ignoring the possibility that it is also grounded on the constitutional right to privacy.

This is not to say that the decision of the Washington Court of Appeals in *Compassion* was correct or that the intended import of *Glucksberg* is mistaken. What I do conclude from my critical examination of the lines of reasoning in the opinions of the Courts in these two cases is that the issue of whether the Due Process Clause grounds a constitutional right to physician-assisted suicide remains open. The gaps in Rehnquist's responses to the arguments of Reinhardt may well enable future petitioners to succeed in asserting some such right. However, they would have to meet a very heavy burden of proof because any future court would be constrained in revisiting a decided issue by *stare decisis*, the established policy of our courts to stand by precedent and not to disturb any settled point in the law.

CHAPTER 4

A LEGAL RIGHT TO PHYSICIAN-ASSISTED SUICIDE

Ought there to be a right to physician-assisted suicide under United States law? In this essay I shall argue in the affirmative. I suggest that it should have the same basic structure as the various rights to physician-assisted suicide, constitutional or statutory, that have been claimed in federal and state courts during the past few years. It should be a rights-package, not a single complex right but a set of rights concerning distinct aspects of physician-assisted suicide. Specifically, it should consist of the bilateral liberty-rights to request or not request, to obtain or not obtain, and to use or not use assistance provided by one's physician to commit suicide.

Ideally, it should be a statutory right, more precisely a set of statutory rights, enacted by the legislatures of the several states rather than a constitutional right recognized in federal law.[1] State legislation could and very probably would specify the content and limits of such a right in more detail and with more precision than would be possible in any one or a few decisions of the federal courts. This would provide more guidance and security to both patients and physicians attempting to act in accordance with this right and, at the same time, provide more reliable prevention of abuses by those alleging that their actions are justified by this right. And because state statutes would no doubt differ in various details, the several states would serve as "legal laboratories" to test the actual benefits and harms resulting from the enactment of a right to physician-assisted suicide and to guide improvement of the legislation conferring this right upon qualified patients.[2]

4.1. The Proposed Right

Let me describe briefly the kind of legal right to physician-assisted suicide that I will defend. First, who would be the right-holders? The right would be possessed by all and only adult legally competent patients who are either enduring intolerable unrelievable suffering or who are terminally ill.[3] Many would limit the possession of the right to patients who

[1] In chapter 3, I have argued that the reasoning of Chief Justice Rehnquist in *Washington v. Glucksberg*, 521 U.S. 702 (1997) only partially refutes the arguments of Circuit Court Judge Reinhardt in *Compassion in Dying v. State of Washington*, 79 F. 3rd (9th Cir. 1996). Hence, I believe that were the members of the Supreme Court to become less conservative, it might reconsider its rejection of the constitutional right to physician-assisted suicide in the future. But until it overturns or modifies its holdings in *Glucksberg* and *Vacco v. Quill*, 521 U.S. 793 (1997), there is no such constitutional right.

[2] See *Glucksberg* at 788–789.

[3] I believe that it is safer to begin by conferring this right upon adults only and not minors. Later when there is more empirical evidence of its benefits and harms, I expect that it should be extended to at least emancipated

are both terminally ill and enduring intolerable unrelievable suffering,[4] but to my mind this is too restrictive. I can see no convincing reason why those who are not terminally ill should be required by the law to endure extreme suffering for an even more extended period than those who will soon be released by death or, on the other hand, what value would be protected by denying terminally ill patients the choice of hastening an inevitable death. Again, many would limit possession to those who are enduring suffering resulting from a bodily illness, but why deny relief to those, if any, who are enduring intolerable unrelievable suffering resulting from mental causes? I would agree, however, that even intolerable suffering should not qualify one for a legal right to physician-assisted suicide if the suffering can be rendered tolerable by medical management or eliminated by curing or arresting the illness from which it results.

Second, what would be the defining contents of the legal right to physician-assisted suicide? If it is to serve the purposes and meet the needs of those patients who have claimed some such right in recent court cases, it must be a rights-package consisting of three liberty-rights. On my conception of rights, the essential content of any right is defined by some core position, usually some liberty, claim, power, or immunity.[5] A liberty, in the technical sense in which I am using the term here, is simply the absence of a contrary duty. For example, I have a legal liberty of using my computer at any time of day or night because I have no legal duty to refrain from doing so, but I have no legal liberty to drive while drunk because I have a legal duty not to do so. A bilateral liberty is a liberty to perform *or* not perform some kind of action as one chooses. At the cores of the three rights that constitute the legal right to physician-assisted suicide would be the bilateral liberties to request or not request, to obtain or not obtain, and to use or not use assistance from one's physician to commit suicide. These would be limited to medical assistance, such as lethal medication, from one's primary care physician, the physician who is most likely to have a long-term relationship with the patient. There is no need to rely on one's physician for nonmedical assistance, such as providing a loaded gun or instructions on how best to suffocate oneself with a plastic bag, and one's primary care physician is in the best situation to judge whether one's choice to die is rational in the light of one's basic values.

Third, what would be the most important associated elements in the legal right to physician-assisted suicide? Because no single legal position could confer dominion upon the right-holder in face of some recalcitrant second party, any real right must include more than its defining core.[6] It must be a complex structure including associated legal positions that, if respected, confer freedom and control over that core upon the right-holder. In any effective legal right to physician-assisted suicide, these would include at least the following: (1) The legal duty of others not to coerce or exert undue influence on a qualified patient to request or not request, to obtain or not obtain, or to use or not use medical assistance to commit suicide. Some such legal duty is necessary in order to protect the freedom of the right-holder to exercise the core liberties in her right to physician-assisted

---

and mature minors as Susan Wolf proposes. See, S. Wolf, Facing assisted suicide and euthanasia in children and adolescents. In L. L. Emanuel (ed.), *Regulating How We Die: The Ethical, Medical, and Legal Issues Surrounding Physician-Assisted Suicide* (Cambridge, MA: Harvard University Press, 1998), p. 103.

[4] For example, M. Gunderson and D. J. Mayo, "Restricting Physician-Assisted Death to the Terminally Ill", *Hastings Center Report* 30 (Nov–Dec 2000), 17–23.

[5] C. Wellman, *A Theory of Rights* (Totowa, NJ: Rowman & Allanheld, 1985), pp. 81–95.

[6] *Ibid.* pp. 58–60.

suicide as she chooses. (2) The legal immunity of a qualified patient against having the sale or issuance of any life, health, or accident insurance or any annuity policy, or the rate charged for any such policy, conditioned on or affected by her exercise of any of her core liberties in this right. This legal immunity would also protect, although in a rather different way, the freedom of the right-holder to exercise her core liberties as she chooses. (3) The legal immunity of a qualified patient from having the interpretation of any provision in a contract or will adversely affected by her exercise of any of her core liberties in this right. (4) The bilateral legal liberty of an attending physician to provide or refrain from providing medical assistance to a competent adult patient who is either terminally ill or enduring intolerable unrelievable suffering from an incurable illness and who has requested such assistance. Obviously, the patient's exercise of her liberty-rights to obtain and use assistance from her physician to commit suicide would be greatly hampered by any legal duty of her physician not to provide such assistance. The patient's legal right to physician-assisted suicide can be reliably effective only if her physician is at liberty to assist her to end her life. At the same time, it is unnecessary and undesirable to impose a legal duty upon any physician to assist a patient to commit suicide if this would violate his conscience or go against his best medical judgment. Hence, this associated element should be a bilateral legal liberty to provide or not provide medical assistance as the attending physician chooses. (5) The legal duty of an attending physician who refuses to provide medical assistance to a qualified patient who has properly requested it to refer that patient to another physician. Although the attending physician ought not to be legally required to provide assistance to any patient intending to kill herself, neither ought he to be in a position to obstruct the patient's exercise of her legal right to physician-assisted suicide. No doubt any statutory right to physician-assisted suicide would also include other associated elements, but an illuminating description of the kind of right I propose need not take the form of or be as detailed as an enacted statute.

4.2. Reasons to Enact

The standard arguments in favor of a legal right to physician-assisted suicide are hardly news. It is frequently asserted that there ought to be a right to physician-assisted suicide under United States law in order to enable qualified patients to avoid unnecessary suffering, to enable qualified patients to die with dignity, and to respect those patients' right to autonomy or self-determination. Although I believe that these three arguments point in the right directions, they stand in need of a more precise formulation and fundamental justification than is usual. For one thing, it is essential to distinguish between reasoning from authoritative legal sources to conclusions about what the law is and reasoning from moral premises to what the law ought to be. "The Philosophers' Brief" purports to consist of purely legal reasoning. "These cases do not invite or require the Court to make moral, ethical, or religious judgments about how people should approach or confront their death or about when it is ethically appropriate to hasten one's own death or ask others for help in doing so."[7] But in his introduction to it, Ronald Dworkin explains that its reasoning is

[7] "Assisted Suicide: The Philosophers' Brief", *New York Review of Books*, 27 March 1997, p. 43.

both moral and legal. "First, it defines a very general moral and constitutional principle—that every competent person has the right to make momentous personal decisions which invoke fundamental religious or philosophical convictions about life's value for himself."[8] Moreover, one should distinguish carefully between the principles of individual morality and the moral principles concerning public morality and legal institutions. Finally, those who have advanced these standard arguments have done far too little to justify the moral principles to which they appeal. Let us, therefore, re-examine each of these three arguments.

First, there ought to be a legal right to physician-assisted suicide in order to enable qualified patients to avoid unnecessary suffering.[9] Some patients enduring intolerable unrelievable suffering as well as some terminally ill patients who are enduring lesser but still severe suffering need this legal right, for they cannot escape from their suffering without it. Those who will die within hours or a very few days will soon obtain relief without taking any action, and those who are on life-prolonging intensive care can often end their lives simply by refusing continued treatment. But others are condemned to continuing severe suffering by any legal system that confers no legal right to physician-assisted suicide. To be sure, a few might be able to commit suicide without assistance and a few others might be able to find persons willing to violate the law in order to rescue them from their distress. But most patients would be unwilling to commit suicide under those conditions, for attempts to commit suicide often fail, leaving one in an even worse condition than before, and obtaining or using assistance from another to commit suicide could expose that person to legal sanctions.

The law ought to promote the well-being of the citizens when it can do so effectively and without serious social costs, and above all ought not to harm those subject to it unless this is necessary to prevent even greater harms. Any law that forces patients to endure avoidable suffering harms those patients, and any legal right that enables them to escape from suffering is, other things being equal, in the patients' best interests. A legal right to physician-assisted suicide creates an exception to those laws that otherwise would prevent or hinder qualified patients from escaping from severe suffering. Its core liberties would permit qualified patients to commit suicide with the assistance of their physicians, and its associated elements would go a long way toward enabling them to exercise the liberties to request, obtain, and use medical assistance from their physicians if, but only if, they choose to do so. Therefore, there ought to be a legal right to physician-assisted suicide unless such a right would bring with it very serious social costs or great harms.

Second, there ought to be a legal right to physician-assisted suicide to enable qualified patients to die with dignity.[10] The concept of death with dignity is as profoundly important as it is obscure. The *Oxford English Dictionary* defines dignity as "the quality of being worthy or honourable" and reminds us that historically it has been persons of high estate or social rank who have been thought to be honorable. But Immanuel Kant maintained

---

[8] *Ibid.*, p. 41.

[9] D. W. Brock, "Physician-assisted suicide is sometimes morally justified", *Physician-Assisted Suicide*, R. F. Weir (ed.), (Bloomington: Indiana University Press, 1997), pp. 90, 96. See Also *Quill v. Vacco*, 80 F. 3rd 716 (2nd Cir. 1996) and *Compassion* at 814.

[10] M. Pabst Battin, *The Least Worst Death* (New York: Oxford University Press, 1994), pp. 280–284, and R. Dworkin, *Life's Dominion* (New York: Alfred A. Knopf, 1993), pp. 238–241. See also *Compassion* at 794 and 812 and *Glucksberg* at 779, 790–791.

that there is something in human nature, irrespective of social status, that commands our respect.[11] He believed that this is moral agency and that it confers dignity or inherent worth upon all human beings. Some such thought lies behind the demand for death with dignity.

Human beings share with many nonhuman animals the capacity to feel bodily pleasure and pain, perhaps even to experience happiness and to suffer mental distress. This calls for our concern and compassion, but not our moral respect. I agree with Kant that there is also something in normal human beings that commands our respect and constitutes an essentially human dignity. Although it consists primarily in our practical rationality and capacity for moral choice and action, as Kant held, I believe that it also includes other human capacities such as our imagination, creativity, the ability to communicate and interact with others, and sympathy or an awareness of and concern for the well-being or adversity of others. If this is so, then one's death lacks dignity when the process of dying has destroyed or degraded those essentially human capacities that command our respect. And death can be an indignity when the process of dying flagrantly reveals the gross deterioration of the capacities, such as self-control or the ability to interact meaningfully with others, that constitute one's dignity.

Why does it matter whether or not one dies with dignity? It is important to the patient because when and how one dies profoundly affects the meaning of one's death and, thus, the shape and significance of one's life. One's life is a biography experienced as a drama with a beginning, a middle, and an end such that the intrinsic value of each part is determined much more by one's awareness of its significance for the whole than by its felt pleasantness or painfulness. The awareness that one will die without dignity can undermine one's self-respect and cause one to devalue one's life. The loss of a patient's dignity also affects how others remember her and reduces, at least to some degree, their respect for her. This is an injury to the patient, who must now expect to be remembered less fondly and with less respect than she would wish, and a misfortune to friends and family members, who are condemned to live on with distressful memories of the death of their loved one.

The law ought not to harm patients and those who love them by denying qualified patients the opportunity to die with dignity. Someone who is terminally ill ought not to be forced to submit passively to a death beyond her control, but ought to be able to exercise her moral agency by deciding whether or not to end her life at a time and in a manner of her choosing. Similarly, someone who is enduring intolerable suffering ought to be free to end her life before her human capacities are irreparably damaged either by her suffering or by the illness causing her to suffer. Therefore, there ought to be a legal right to physician-assisted suicide to enable these patients to die with dignity.

The third standard argument is that there ought to be a legal right to physician-assisted suicide in order to respect the qualified patient's moral right to autonomy or self-determination.[12] Although I find this argument very plausible, I hesitate to accept it at face value. I have not been able to find in the literatures of moral philosophy or biomedical

---

[11] I. Kant, *Groundwork of the Metaphysica of Morals*, H. J. Paton (trans.) (New York: Harper and Row, 1964).

[12] Brock, "Physician-Assisted Suicide Is Sometimes Morally Justified", pp. 89–90. See also *People v. Kervorkian*, 527 N.W. 2nd 714 (Mich. 1994) at 727, *Quill* at 727 (Quoting *Rivers v. Katz*, 495 N.E. 2nd 337), *Glucksberg* at 724 and *Krischer v. McIver*, 697 So. 2nd 97 (Fla. 1997) at 111 & 114.

ethics any clear definition of the content of the alleged moral right to autonomy, much less any convincing explanation of its grounds. For example, Ronald Dworkin defines it as the right of every competent person "to make momentous personal decisions which invoke fundamental religious or philosophical convictions about life's value for himself."[13] One can understand why he would define the constitutional right to autonomy in terms that echo the language of *Planned Parenthood v. Casey*,[14] but very few moral philosophers conceive of the right to self-determination in these terms. I also agree with Chief Justice Rehnquist that we need a more careful formulation[15] of the content of the right to autonomy in order to understand precisely how it bears on the legal and moral issues concerning physician-assisted suicide. Even more urgent is the need to identify the moral reasons sufficient to establish the existence of any alleged moral right to autonomy. Dworkin suggests that it is grounded on some sort of a fundamental moral right to dignity or respect.[16] However, he neither defines the content of this fundamental moral right clearly nor gives any reasons to show that it really exists. Other moral philosophers have not, to my knowledge, filled these gaps in moral theory. Hence, I prefer to rest my case on two more simple considerations that might well underlie a right to autonomy, the moral duty not to intrude into the life of another and the moral duty to respect the rational agency of others.

(1) Everyone has a moral duty not to intrude into the life of another. One intrudes in the literal sense when one invades someone's home without her permission, but what makes this morally wrong is the way in which it intrudes into the life of the person living there by causing her distress and disrupting her activities. There are two morally relevant aspects of intrusive action. It disrupts the ongoing experience and action of the person subjected to it, and it forcefully enters that life from the outside. As disruptive it is experienced as worrisome and frustrating, and it often damages one or more projects of the victim. As an invasion by some alien will it is resented and alienates the person subjected to interference from the intruder. Hence, it is morally wrong to intrude into the life of another.

The sort of intrusion that is relevant here is the restraint imposed on the patient by laws that prevent or hinder her from acting on her choices. The early right-to-die cases, such as *Quinlan*,[17] argued that the patient has a right to refuse life-sustaining treatment that is intrusive or invasive. That legal prohibitions can also be intrusive was recognized in *Roe v. Wade*[18] and other early abortion cases that appealed to the patient's right to privacy. Laws prohibiting physician-assisted suicide intrude into the life of a patient who chooses to die with the assistance of her physician by preventing or hindering her from acting on her decision and thus pursuing one of her vital projects. Moreover, this intervention is correctly experienced as a disruptive intrusion into one's life by an external and hindering force, an alien will.

---

[13] R. Dworkin, Introduction to "Assisted Suicide: The Philosophers' Brief", p. 41.

[14] *Planned Parenthood of Southeastern Pennsylvania v. Casey*, 505 U.S. 833.

[15] *Glucksberg* at 722.

[16] Dworkin, *Life's Dominion*, pp. 238–239. See also Ronald Dworkin, *Taking Rights Seriously* (Cambridge, MA: Harvard University Press, 1977), p. 198.

[17] *In the Matter of Quinlan*, 70 N.J. 10.

[18] 410 U.S. 113.

The reasons that ground the moral duty of individuals not to intrude into the life of another are equally relevant to an analogous political norm. The law ought not to intrude into the momentous private decisions of those subject to it. Many minor legal restrictions are justified as conducive to important public goals, and more serious legal prohibitions are justified when they are necessary to protect others from serious harm. But a momentous private decision is one that fundamentally affects the life of one individual for better or worse and such that no one else has a sufficient reason to prevent one from acting on it. The law ought not to interfere in any such decision or hinder an individual from acting on it unless this is necessary to protect some important state interest. Presumably, the decision of a terminally ill patient or a patient enduring unrelievable intolerable suffering to commit physician-assisted suicide is a momentous private decision. Therefore, I infer that there probably ought to be a legal right to physician-assisted suicide.

(2) Everyone has a moral duty to respect the rational agency of others. Respect in the relevant sense is a deferential esteem manifested by a disposition to yield to the choices and actions of another person. Esteem is more than a feeling of respect; it also involves a judgment that the rational agency of another is worthy of deference. And what the duty of respect requires is not some inner feeling plus judgment, but conduct that is overtly deferential by allowing another person to act as she has decided to act.

Why is there any moral duty to respect the rational agency of others? Rational agency involves setting one's goals, selecting ways of achieving them, and integrating ends and means into a more or less coherent life. But this becomes impossible, or at best very difficult, if others do not defer to one's decisions and yield to one's actions. It is projects that enable one to realize one's most valued goals, and one's projects give the coherence and meaning to one's life that make it more than a mere series of trivial satisfactions. Hence, to disrespect the rational agency of another by interfering with her action is to reduce her control over her life and thereby to threaten what matters most to the value of her life. In addition, the respect of others is a necessary condition for sustaining self-respect. Thus, if others fail to respect one's rational agency, one tends to lose confidence in one's ability to choose and act rationally. This reduces or destroys the initiative, creativity, endeavor, persistence, and self-reliance required to live a rewarding life and to contribute to the lives of others. These harmful consequences of failing to respect the rational agency of others ground our moral duty of respect.

They also support a similar political norm. The law ought not to limit or deny individual liberty except when necessary to protect important state interests. Obviously any laws that prohibit or seriously hinder a patient from committing suicide with the assistance of her physician limit the liberty of the individual patient and her physician. It is hard to see what important state interest would make it necessary to deny individuals who are terminally ill or are enduring intolerable unrelievable suffering liberty in this manner. Therefore, there very probably ought to be a legal right to physician-assisted suicide.

Thus, an examination of the three standard arguments in favor of a legal right to physician-assisted suicide reveals four reasons for state legislatures to enact statutes that would confer this right upon adult legally competent patients who are terminally ill or enduring intolerable unrelievable suffering resulting from an incurable illness. The several states ought to introduce and maintain a legal right to physician-assisted suicide in order to enable qualified patients to avoid unnecessary suffering, to enable them to die with dignity, to avoid intruding into their lives, and to avoid denying or limiting individual liberty unnecessarily.

4.3. Arguments Against Enactment

The four reasons that I have explained briefly in the previous section strongly support the conclusion that the several states ought to enact legislation that would confer a right to physician-assisted suicide upon qualified patients, but this is only a *prima facie* conclusion. They show that there ought to be a right to physician-assisted suicide under United States law *unless* it is necessary to deny this right to adult legally competent patients who are terminally ill or enduring intolerable unrelievable suffering in order to prevent some great social harm or to promote an important state interest. Let us examine the most plausible arguments of those who believe that this is so.

(1) It is necessary to prohibit physician-assisted suicide in order to promote the important state interest in the preservation of human life.[19] Although I readily grant that every state does have an important interest in the preservation of human life, I doubt that this rules out the right to physician-assisted suicide. Why do states have an interest in preserving human lives? One reason is that there is a human right to life and the state ought to protect the human rights of its subjects. So far, so good, but how does this apply to physician-assisted suicide? Many argue that suicide is self-murder and thus any patient who exercised her right to physician-assisted suicide would violate her own right to life. It is hard to know how to assess this reasoning because the human right to life is often asserted but seldom defined. I believe that the human right to life is a rights-package including at least the individual's claim-right not to be killed by others.[20] Correlative to this claim is the duty of others not to kill one. But because the notion of a moral claim against oneself makes no sense, killing oneself is not an analogous violation of one's own human right to life. However, I also believe that moral agents have a general duty, a duty under normal circumstances, not to commit suicide. But at least in the case of patients suffering intolerable unrelievable suffering, this duty is undermined by excessive sacrifice. There are limits to moral obligation. Although a firefighter has a duty to enter a burning building to rescue a resident, she does not have any duty to do so when the building is about to collapse and any attempt to rescue would mean certain death. Similarly, patients have no moral duty to refrain from suicide when this would mean intolerable suffering.

Still, one's physician could easily refrain from assisting a patient to commit suicide and arguably ought to do so because assisting a patient to commit suicide amounts to killing that patient, thus violating her claim-right not to be killed by another. But the physician does not kill the patient by providing her with the means to kill herself. If the patient chooses to use these means to commit suicide, it is the patient who does the killing. Nor is the physician an accomplice in an immoral killing, because the qualified patient has no moral duty not to commit suicide. In any event, the patient will have waived any moral rights holding against the physician by requesting her assistance. Granted that the human right to life is inalienable, waiving a right is not alienating it. The patient retains her right to life, as evidenced by the fact that if some misguided stranger were to kill the patient, he would be violating her right to life.

---

[19] See *Compassion* at 817–820 and *Glucksberg* at 728–730.

[20] C. Wellman, The inalienable right to life and the durable power of attorney. *An Approach to Rights: Studies in the Philosophy of Law and Morals* (Dordrecht: Kluwer Academic Publishers, 1997), pp. 245–248.

Another reason why states have an important interest in preserving human life is that most individuals are capable of contributing to society, if not by public service at least by being productive members of society and useful to their family and friends. However, patients who are terminally ill or enduring intolerable irremediable suffering are usually no longer capable of serving others in the usual ways. Although the brave acceptance of suffering can serve as an admirable moral example to loved ones, those desperate enough to choose physician-assisted suicide may submit to but do not accept their unnecessarily prolonged agony. More often than not they are a burden, both financially and emotionally, upon their families and heavy consumers of medical resources that could much more usefully be allocated to other patients. Therefore, the state interest in preventing the loss of lives capable of contributing to society is not an important consideration regarding any right to physician-assisted suicide.

A third reason why the state has an interest in the preservation of human life is that far too many lives are lost prematurely. In particular, suicide is a serious social problem because persons who are mentally ill or temporarily suffering from deep depression commit suicide even though they could have lived long, productive, and reasonably happy lives. For this reason state statutes permitting suicidal persons to be involuntarily confined for a limited period while they obtain psychiatric or medical treatment are justified. But the persons qualified to possess the legal right to physician-assisted suicide I have described and who might exercise it to commit suicide would not die prematurely and are not capable of living long and worthwhile lives. Hence, the state's general interest in preventing suicide does not apply to them. I conclude that the very important state interest in the preservation of human life is not a sufficient reason why there ought not to be a legal right to physician-assisted suicide.

(2) It is necessary to prohibit physician-assisted suicide in order to promote the important state interest in maintaining the ethics and integrity of the medical profession.[21] Even those who accept the need for assisted suicide sometimes object to permitting a physician to provide assistance. They argue that this would be inconsistent with the moral obligations of the physician as physician. Each profession is defined by some special purpose. The purpose of teachers is to educate their students; the purpose of lawyers is to serve the legal interests of their clients; and the purpose of physicians is to promote the health of their patients. This defining purpose imposes on the physician the duties to cure illness, to remove or mitigate disability, and at least to preserve life. It would be contrary to the ethical standards of the medical profession to permit physicians to destroy human life, either directly by killing a patient or indirectly by assisting a patient to kill himself.

I do not believe that one can define the role of the physician so simply. What physicians can and ought to do in any society is only partly defined by the traditional purposes of the profession. Also, one traditional duty of the physician has been to relieve the suffering of her patient even when to do so may threaten her life. It is not a big step from administering dangerously large doses of pain-reducing morphine to providing lethal medication to a patient who chooses to end his suffering. Moreover, how a physician ought to treat her patients changes as new medical technologies become available and the medical problems

[21] See *Compassion* at 827–830 and *Glucksberg* at 731.

of her patients change. The ability of modem medicine to extend human life is a blessing to many but a curse to others, in particular to some patients who are terminally ill or enduring intolerable unrelievable suffering. Surely medical practitioners have some moral responsibility for the avoidable suffering they impose on those patients. It seems to me that physicians ought to be legally permitted to discharge this responsibility by assisting qualified patients who choose to end their lives.

To be sure, the state has a very important interest in the integrity of the medical profession. Because medicine is an invaluable social institution, society must preserve the public trust in medical practitioners. If the law were to permit physicians to hasten the deaths of their patients, some argue that those with the most serious medical problems would be reluctant to put their lives in the hands of the medical establishment and thus would not receive the care they need, and the families of patients who might be tempted to commit suicide prematurely may refuse to cooperate with physicians they have come to distrust. It is essential that physicians not only act morally but also that they be perceived to be morally impeccable.

I agree that the law ought to protect the integrity and public image of the medical profession, but I deny that this is best accomplished by prohibiting physician-assisted suicide. Rather, the reasons to enact statutes conferring a right to physician-assisted suicide to which I have appealed in the previous section of this essay should be explained clearly and repeatedly to the public. Indeed, the standard arguments for the right to physician-assisted suicide are already familiar to large segments of the public and are producing political pressure to enact a legal right to physician-assisted suicide. The probability that patients will deny themselves needed medical treatment because they fear for their lives is greatly exaggerated. After all, their physicians would be permitted to provide assistance only if requested, and it remains up to the patient whether or not to use any lethal medication thus obtained. It is much more likely that patients will be reluctant to admit themselves to hospitals for fear that they will be unable to escape the clutches of physicians legally required to keep them alive no matter how pointless life has become for them and how great the agony they must endure.

(3) There ought not to be any legal right to physician-assisted suicide because any such right would be seriously abused.[22] For one thing, it would cause right-holders who would prefer to live to be pressured, even coerced, into committing physician-assisted suicide against their wills. Family members under emotional and financial strain would be sorely tempted to talk a patient into exercising his right to commit physician-assisted suicide. Even the most loving and conscientious family members might well rationalize that death would be best for the patient. Health care institutions vigorously pursuing cost containment and mindful of the disproportionate amount of medical resources allocated to patients during the last year of their lives might refuse to finance medical care desired by those qualified to exercise the right to physician-assisted suicide. Even physicians, many of whom are dedicated to curing illness and overcoming disability and are reluctant to care for patients they think of as hopeless, might unintentionally exert undue influence by the way they present their diagnoses and explain the prognosis to patients qualified to commit physician-assisted suicide.

---

[22] See *Lee v. State of Oregon*, 891 F. Supp. 1429 (D.Or. 1995) at 1438 and *Krischer* at 101–102.

Also any legal right to physician-assisted suicide would result in discrimination against the vulnerable.[23] Our stereotypes condition us to undervalue the lives of several categories of persons, especially the poor, the elderly, the mentally ill, and the disabled or handicapped. We imagine that, unfortunately, none of them can live as productive, rewarding, and enjoyable lives as normal persons. It might seem merciful to inform them of their right to commit physician-assisted suicide and encourage them to exercise this right. Members of these groups are less able than others to resist such pressures and would in practice end their lives prematurely in disproportionate numbers.

Finally, the existence of a legal right to physician-assisted suicide would lead physicians to provide lethal medication to unqualified patients. It would in practice be impossible to confine physician-assisted suicide to legally competent patients who are terminally ill or enduring intolerable unrelievable suffering. The concept of terminal illness involves a prediction that, given the variability of individual cases, can never be certain, and the concept of suffering is so subjective that there are no objective measures of the degree of suffering. Hence, attending physicians would often provide lethal medication to patients whose illness could be at least partially reversed and to demanding patients whose suffering is not intolerable, perhaps not even especially severe. Presumably, only a patient who is competent to make a rational decision would possess any right to physician-assisted suicide. But patients who are seriously ill or enduring suffering tend to become depressed, and the major cause of premature suicide is depression. Physicians are not professionally qualified to distinguish between mild depression that leaves a patient capable of making a rational decision and a deeper depression that renders a patient legally incompetent. Hence, once again physicians would often provide assistance to suicidal patients who do not possess any legal right to physician-assisted suicide.

To my mind, and that of many others, the danger of abuse is the strongest argument against having any legal right to physician-assisted suicide under United States law. Each of these abuses is possible and highly undesirable. The question, and it is an empirical question, is how often they would in fact occur. To date there is insufficient empirical evidence to answer this question with any confidence. Interpretations of the studies of physician-assisted suicide in The Netherlands, where it has been practiced for several years, reach conflicting conclusions.[24] Moreover, the relevance of these studies of medical practices in a jurisdiction that permits euthanasia as well as physician-assisted suicide to the proposal to introduce only physician-assisted suicide remains in doubt. Here in our country, it is too early to draw any reliable conclusions about the dangers of the Oregon Death With Dignity Act. However, some of the evidence is reassuring. Fewer patients have requested lethal medication from their physicians than was anticipated, and a considerable number of patients who have obtained such medication have chosen not to use it to commit suicide but have died of natural causes.[25] One advantage of introducing this right by legislation in the several states, rather than by a decision of the United States Supreme Court, is that the states would enact somewhat different protective regulations. This would over time provide empirical evidence of whether most serious abuses of the

[23] See *Glucksberg* at 731–732.
[24] See *Glucksberg* at 785–786.
[25] See the Annual Reports on Oregon's Death with Dignity Act available on the Internet at http://www.ohd.hr.state.or.us/pas/pas,htm.

legal right to physician-assisted suicide can be prevented and, if so, how to do so most effectively.

My semieducated guess is that the kinds of abuses mentioned above can be held to an acceptable minimum.[26] Some guidance on how this might be accomplished can be found in "A Model State Act to Authorize and Regulate Physician-Assisted Suicide" proposed by Charles H. Baron and others.[27] In addition to specifying the conditions necessary for a patient to possess the right to physician-assisted suicide, this model statute specifies that a physician is permitted to provide lethal medication only if the patient has made a request that is not the result of clinical depression or any other mental illness, represents the patient's reasoned choice, has been made free of undue influence, and has been repeated without self-contradiction on two separate occasions at least 14 days apart.[28] Moreover, before providing medical means of suicide, the physician must offer to the patient all available medical care, including hospice care, secure a written opinion from a consulting physician that the patient is suffering from a terminal illness or an intractable and unbearable illness, and secure a written opinion from a psychiatrist, clinical psychologist, or psychiatric social worker that the patient's request is not the result of any distortion of the patient's judgment due to clinical depression or any other mental illness.[29] There are also requirements concerning documentation and reporting to enable medical institutions and state authorities to oversee and enforce these and other regulations. It is premature to judge that some such set of regulations cannot effectively protect patients from serious abuses of the legal right to physician-assisted suicide.

(4) There ought not to be a legal right to physician-assisted suicide because this would be the first step toward voluntary than nonvoluntary and even involuntary euthanasia.[30] Besides the difficulty of drawing the line between qualified and unqualified patients who request assistance to enable them to commit suicide, there is the difficulty of holding the line between assisting a patient to kill herself and killing a patient who requests that her life be ended. Once it becomes accepted that physicians are permitted, perhaps even ought, to assist some of their patients to die, the line between assisting a patient by prescribing lethal medication for the patient to use and assisting more directly by injecting it into the patient becomes much less important. Indeed, these are merely different methods of providing lethal medication to one's patients. Physicians will often cross that line when a competent patient who is terminally ill or enduring intolerable unrelievable suffering is physically or psychologically incapable of using assistance to commit suicide. Then the permissibility of nonvoluntary euthanasia, the mercy killing of a patient who has not requested assistance in dying, seems to follow legally because a request by a surrogate decision-maker counts legally as a request by an incompetent patient. Thus, what began as a legal right to enable qualified competent adults to die with dignity will lead to a practice

[26] More educated opinions that agree with mine are found in Brock, "Physician-Assisted Suicide Is Sometimes Morally Justified," p. 540; *Kevorkian* at 751; *Quill* at 730; and *Compassion* at 831.

[27] C. H. Baron et al., *Harvard Journal on Legislation* 33 (1996), 1–34; see also, R. Cohen-Almagor and M. G. Hartman, "The Oregon Death With Dignity Act: Review and Proposals for Improvement", *Journal of Legislation* 27 (2001), 293–298.

[28] *Ibid.*, p. 27.

[29] *Ibid.*, p. 27, 29.

[30] See *Compassion* at 830–832 and *Glucksberg* at 732–735 and 782–785.

of killing children, the mentally ill, and those suffering from senile dementia without their consent or even against their wishes.

This is a slippery slope argument. It assumes that legally we are now standing on relatively flat ground protected from sliding into harmful and immoral practices by the legal principle that physicians must always preserve human life *and* that if we take the first step down the slope by permitting physicians to assist a few patients to kill themselves we will be unable to retain our legal prohibitions against physicians killing more and more of their patients. I do not find this argument persuasive. For one thing, we are by no means standing on legally level ground. It is legally permissible, and generally accepted as morally justified, for physicians to withhold or even withdraw life-preserving medical treatment at the request of a patient or a qualified surrogate decision-maker. Moreover, there is little evidence that this first long step down the slope from the unconditional legal protection of the life of every patient has caused any significant increase in euthanasia. The slope is not nearly as slippery as the argument assumes. Whether the next step of introducing a legal right to physician-assisted suicide will take us onto more slippery ground depends on how clearly one can distinguish each successive step leading down to involuntary euthanasia. If there is no sharp and important distinction between one step and the next, then there is no strong reason not to take that additional step; if there is a sharp and important distinction, then this does provide a strong reason to halt legal reform at this point. Why assume that political pressures and legal processes are so irrational that we will be unable to prevent our legal system from descending into immorality?

(5) There ought not to be a legal right to physician-assisted suicide because this would change the social scene in a way that would be a spiritual disaster.[31] This is a much more general and fundamental concern than predictions of specific abuses of the right or of the slide from that right to more objectionable changes in our legal system, for it envisions widespread side-effects on the expectations, attitudes, motivations, and moral principles that enrich our interpersonal relationships and sustain the social fabric of our society. Some moral philosophers and jurists argue that a legal right to physician-assisted suicide would undermine the expectation of most people that they will be cared for if they become old or ill. It would even become expected that someone who needs a great deal of care should demand that the doctor help one to die. It would exacerbate our excessive reliance on medical technology at the expense of the more natural human associations, especially the family, on which we need to rely when medicine can no longer cure our illness or alleviate our suffering. It would reinforce societal indifference and even our prejudicial undervaluation of the lives of the elderly, the terminally ill, and the disabled—the most vulnerable members of our society. While a humane society provides support of every kind to those who are burdened in order that they may live, introduction of a legal right to physician-assisted suicide would diminish if not destroy society's respect for the value of every human life no matter what its condition.

I do not believe that the introduction of a legal right to physician-assisted suicide would in fact result in anything like the spiritual disaster that Philippa Foot and others fear. There is no reason that it should damage our expectation that we will be cared for when

---

[31] See P. Foot, "Euthanasia", *Philosophy and Public Affairs* 6 (1977), 85–112; L. R. Kass, "Is There a Right to Die?" *Hastings Center Report* 23 (1993), 34–43; *Kevorkian*, at 729 n. 45 quoting *Guardianship of Jane Doe*, 583 N.E. 2nd 1263 (1992) and *Glucksberg* at 732.

we become old or ill. Rather, it will reflect and reinforce a deeper and more compassionate understanding of proper medical care. Those who truly care about the welfare of the patient should not insist upon the prolongation of her physiological and psychological existence against her will, but should recognize that proper medical care includes enabling those who are terminally ill to die with dignity and those who are enduring intolerable suffering to end their agony if they so choose. Nor would legally permitting these categories of patients the possibility of assistance in committing suicide if, but only if, they so choose exacerbate our excessive reliance on medical technology. Permitting patients to refuse life-prolonging medical treatment has not had this result. It is the progress of medical science and the development of new medical technologies together with the virtual disappearance of the extended family that are forcing increasing numbers of patients to die in hospitals, hospices, or nursing homes rather than in their family homes. Denying them the legal right to physician-assisted suicide will do little or nothing to slow the advance of medical technology or to restore the more natural human associations of a bygone era. A legal right to physician-assisted suicide would not devalue the lives of the most vulnerable members of our society; it would recognize that society ought to accept their own evaluations of their lives and not impose external moralistic constraints upon how they are permitted to live and die. Finally, although a humane society ought to provide support of every kind to those who are burdened with illness or suffering and wish to live, introducing a legal right to physician-assisted suicide will not undermine our respect for each and every human right. A true respect for a genuinely human life must include a respect for the human capacity to make rational choices regarding how and how long one should live. To equate respect for life with the mere prolongation of living is to reduce the life of a person with the normal capacities and potentialities of a human being to the mere biological functioning of an organism.

(6) There ought not to be a legal right to physician-assisted suicide because it would expose patients to unnecessary risk, the ultimate risk of premature or undesired death.[32] To be sure, the dangers many fear are speculative and not grounded on conclusive empirical evidence, but neither is there any solid evidence to demonstrate that they are unreal. Even if the optimists are correct in predicting that a right to physician-assisted suicide would not be very widely abused or greatly increase the incidence of euthanasia in our society, no set of regulations can eliminate all abuses or totally inhibit the temptation to practice mercy killing. If permitting these grave risks were necessary in order to avoid some greater evil, this would be justified, but many argue that the introduction of a legal right to physician-assisted suicide into United States law is not necessary. While such a right would put large numbers of patients at risk, very few would benefit from this right. Many patients who are terminally ill or suffering intolerable unrelievable suffering are receiving life-sustaining medical treatment. All they or their surrogate decision-makers need to do is to withdraw their consent and they will soon die from natural causes. Many others would not be qualified to exercise any legal right to physician-assisted suicide because they are too young or mentally ill or emotionally disturbed or elderly persons with senile dementia and thus not legally competent. For those very few who would be entitled to practice physician-assisted suicide, there are preferable alternatives. Although pain management

[32] See *Kevorkian* at 742–743.

is not practiced as effectively as possible in most hospitals, modern medicine can now relieve almost all bodily pain, and psychological suffering can also be controlled with new medications. Hospice care can and should be made available to all terminally ill patients so that they are not condemned to endure the dying process in an impersonal hospital setting or in an even more dismal nursing home. In those highly exceptional cases in which a legal right to physician-assisted suicide might be useful, the patient may well be able to find a caring physician willing to assist her in dying without regard to the letter of the law.

It is true that some, one does not know how many, patients are fortunate enough to find physicians willing to ignore the law and assist them to commit suicide. But presumably this clandestine medical practice is more dangerous than any more open practice subject to the protective legal regulations that would be included in any statute conferring a legal right to physician-assisted suicide. It is also unfair to threaten conscientious physicians with legal sanctions for enabling their patients to exercise their moral right to physician-assisted suicide, a right I defend in the next essay. Equally unfair is the present situation, in which the opportunity to exercise this moral right is enjoyed mainly by affluent patients but usually denied to members of disadvantaged groups who cannot afford the ongoing medical care needed to establish a relationship with one's physician that is close enough to motivate her to violate the prohibition against assisted suicide. A legal right to physician-assisted suicide would provide more equal protection of the law to all patients. Finally, there is no doubt that current medical practice could be improved in ways that would provide preferable alternatives to many, not all, patients who are terminally ill or enduring intolerable suffering. But it is cruel to deny physician-assisted suicide to those patients who need it now while those alternatives are not yet available. Granted that a legal right to physician-assisted suicide would expose some patients to danger, this is not an unnecessary risk.

(7) Enacting a legal right to physician-assisted suicide is undesirable because it would reduce the need to introduce better ways of caring for the dying and improved techniques for relieving the suffering of most patients.[33] Granted that both are often inadequate today, considerable progress has been made recently in both respects. Many argue that a right to physician-assisted suicide would retard, perhaps reverse, this progress by reducing its urgency. Because terminally ill patients could exercise this right in order to avoid an undignified death, there would be less need to use scarce medical resources to maintain high-quality hospice care. And because patients enduring intolerable suffering would also have this option, there would be less need to provide more effective pain management or more adequate psychological support services in hospitals or nursing homes. Thus, the introduction of a legal right to physician-assisted suicide would result in premature or unnecessary death for patients who would be much better served by sustaining our progress in the introduction of preferable alternatives.

Once more I insist that it is cruel to deny physician-assisted suicide to those patients who are terminally ill or enduring intolerable suffering, be they many or few, and for whom adequate hospice care and the relief of their agony are not yet available. It is unjust to sacrifice them for the sake of future patients who would benefit by desirable

---

[33] D. C. Thomasma, "When Physicians Choose to Participate in the Death of Their Patients", *Journal of Law, Medicine and Ethics* 24 (1996), 183–197, and D. Orentlicher, "The Legalization of Physician Assisted Suicide: A very Modest Revolution", *Boston College Law Review* 38 (1997), 443–475.

improvements in medical practice. Moreover, our society has no need to choose between enacting a legal right to physician-assisted suicide and continuing the improvement in caring for the dying and relieving suffering. Both reforms, legal and medical, can and should be vigorously pursued. If as a consequence very few patients choose to exercise their right to physician-assisted suicide, so much the better.

I conclude that each of the several states ought to enact a right to physician-assisted suicide. It would be useful for the various state statutes to differ in details, especially regarding protective regulations, in order to obtain empirical evidence concerning the least dangerous and most beneficial formulation. Still, all of the rights conferred should have the same basic structure. They should be possessed by all and only those adult legally competent patients who are either enduring intolerable unrelievable suffering or terminally ill. They should be rights-packages consisting of three liberty-rights—the rights to request or not request, to obtain or not obtain, and to use or not use assistance provided by one's physician to commit suicide. In addition to its defining core bilateral liberty, each of these rights must include associated elements that confer dominion, freedom, and control over this liberty upon the right-holder. Among these will be the legal liberty of the attending physician to provide, subject to protective regulations, requested medical assistance to her patient.

I do not take the arguments against enacting a legal right to physician-assisted suicide lightly. They are serious arguments, both because they concern important public interests and because there is not yet sufficient empirical evidence to predict reliably how much, if at all, any such right would damage those interests. Nevertheless, I have explained why I do not believe that they outweigh the reasons in favor of enacting a legal right to physician-assisted suicide. These are to enable qualified patients to avoid unnecessary suffering, to enable them to die with dignity, to avoid intruding into their lives, and to respect their rational agency. These compelling moral considerations cry out for the reform of United States law to make it less inhumane and more just.

CHAPTER 5

A MORAL RIGHT TO PHYSICIAN-ASSISTED SUICIDE

The debate about the permissibility of physician-assisted suicide continues unabated in our courts, legislatures, and throughout the popular media. Scholarly treatments in legal periodicals and philosophical journals have done more to reveal the complexity and difficulty of the issues involved than to resolve them. The focus of most discussions has been on our legal system, especially on whether there is a constitutional right to physician-assisted suicide or there ought to be such a statutory right. Some advocates have appealed to an alleged moral right to physician-assisted suicide; at the same time, many opponents have denied that any such moral right exists. Unfortunately, no one has defined with precision the right she was asserting or denying and then thoroughly examined the possible grounds of this right. The purpose of this essay will be to complete both tasks.

There are a variety of ways one might interpret the moral right to physician-assisted suicide, but it is neither possible nor desirable to discuss them all here. The interpretation offered in this essay seeks to satisfy two criteria. One ought to define a moral right to physician-assisted suicide in such a way that it is plausible to assert its existence, even if in the end this claim is found to be unjustified; and one ought to define this right so as to render it directly relevant to our contemporary political and legal debates. Although several legal rights to physician-assisted suicide have been at issue in recent court cases, they all share a common complex structure. Each has been a rights-package, not a single right but a set of different but essentially related rights concerning some specific activity or subject matter. They have consisted of the three legal liberty-rights of requesting or not requesting, obtaining or not obtaining, and using or not using assistance from one's attending physician to commit suicide. Hence, let us define the moral right to physician-assisted suicide as a rights-package consisting of three analogous moral rights. Because these liberty-rights need not imply any moral duty of the patient's physician to provide assistance should she consider the request immoral or even seriously misguided, the assertion of this rights-package is much more plausible than the assertion of any moral claim-right to physician-assisted suicide that would impose this correlative duty. Does a moral right to physician-assisted suicide, as defined above, really exist?

5.1. The Liberty to Use Assistance

The most controversial of these liberty-rights is the moral right to use or not use assistance from one's physician to commit suicide because to assert this right implies that some patients do have a moral liberty-right to commit suicide. But does anyone really have this moral right? Any real right consists of a defining core together with associated positions that confer dominion, consisting of freedom and control, over that core upon the right-holder.[1] The defining core of any right to commit suicide would imply a moral liberty of killing oneself. Let us begin by asking whether some patients do have this moral liberty. The expression "moral liberty" is used here in a technical sense analogous to one of the fundamental legal conceptions identified by Wesley Newcomb Hohfeld.[2] To say that someone has a liberty to perform some action is simply to say that she does not have any duty to refrain from doing so. Hence, our question becomes whether everyone has a moral duty not to kill oneself. This in turn becomes the question of whether there are any moral duty-imposing reasons that ground a duty not to commit suicide. Moral reasons are dual-aspect, essentially social practical reasons.[3] As dual-aspect reasons they are reasons both for agents to act or refrain from acting in some manner *and* for others to react in some way to any agent who acts in conformity with or contrary to these reasons. What distinguishes specifically moral reasons from other kinds of dual-aspect reasons is that they pertain to society in the sense defined by the *Oxford English Dictionary* as "Association with one's fellow men, esp. in a friendly or intimate manner; companionship or fellowship." Thus, they are reasons for action only for anyone living in society with others, and they are reasons for others to react only to agents with whom they associate from time to time. This is because moral reasons involve sociability factors, factors that contribute to or detract from worthwhile stable ongoing interactions among individuals. Duty-imposing moral reasons are one species of moral reasons. They are reasons for any agent to act or refrain from acting in some manner and for those in society with that agent to react negatively if she fails to do so. For example, the fact that John promised Jane that he would give her a ride to work Monday is a reason for John to do so and for others to react negatively to John, at least by disapproving of him, were he to fail to keep his promise. But the moral force of this reason for John presupposes that he interacts with Jane with some regularity, and only others who associate with John in some way have any reason to be concerned by the unreliability he would demonstrate were he to break his promise to Jane.

The question of whether one has a duty not to kill oneself is best approached by asking why it is that one has a moral duty not to kill another person. What specifically moral reasons ground the duty not to commit homicide? Three moral duty-imposing reasons leap to mind.

(1) The act of killing another person violates the victim's moral right to life. The right to life is a rights-package consisting of a number of rights concerning the life of the right-holder.[4] The least controversial of these is the moral claim-right of any normal human being not to be killed by another person. The ground of this moral right is the grievous harm normally caused for the victim by the loss of her life. This harm does not consist in the termination of one's biological life in and of itself, but in the fact that when one ceases to be alive, one's personal history also ceases. The grievous harm lies in the ending of *a life*, the termination of the victim's biography. The value of anyone's life is intrinsic to the conscious living of it; it inheres in both the various experiences that

[1] A. Carl Wellman, *Theory of Rights* (Totowa, NJ: Roman & Allanheld, 1985), pp. 81–96, 161–169.
[2] W. N. Hohfeld, *Fundamental Legal Conceptions* (New Haven: Yale University Press, 1919), pp. 38–39, 42–43.
[3] C. Wellman, *Real Rights* (New York: Oxford University Press, 1995), pp. 39–48.
[4] C. Wellman, *An Approach to Rights* (Dordrecht: Kluwer Academic Publishers, 1997), pp. 245–248.

make it up and the meaning or significance of these evolving experiences to the person having them. The act of killing another typically deprives the victim of all the value she would have experienced were her life to have continued. This great harm is both a reason for one person not to kill another and for those in society with one who commits a homicide to react negatively to an agent who displays such a disregard for the welfare of another.

(2) The action of killing another person also normally harms many of those who associate with the victim—family, friends, and colleagues. Those who feel close to the victim will feel sorrow and grief, often deeply distressing feelings. Some of them may well feel guilty for not having done more to protect the victim from an untimely death. Moreover, the act of killing one person often deprives others of her services. The family loses a breadwinner and/or homemaker, a firm loses a valuable employee, a class no longer has an effective teacher, and a team loses one of its members. Finally, killing a person usually deprives her associates of her companionship, an important value in their lives. These harms are both reasons for an agent not to kill another person and reasons for those in society with anyone who does commit homicide to react negatively to the killer.

(3) Any act of homicide threatens the security of others in the agent's community. Even those who are not close associates of the victim now have some reason to fear for their lives, or at least their physical well-being. It is not merely that a feeling of fear may intrude into their consciousness or a subconscious anxiety destroy their peace of mind. They may curtail worthwhile activities to avoid danger and expend valuable resources to enhance the security of their homes, business establishments, or automobiles. These harms to the wider community are both reasons for an agent not to commit homicide and for the members of a community to react negatively to any agent who has shown himself to be dangerous by killing another person. Here, then, are three specifically moral reasons that ground a moral duty not to kill another person.

Do these or similar reasons also ground a moral duty not to kill oneself? (1) It is difficult to believe that the act of suicide violates the victim's moral right to life. Although any normal human being has a moral claim-right not to be killed by another person, no one can have a moral claim holding against herself. To be sure, one can have moral duties regarding oneself—for example the duty not to mutilate one's body and perhaps the duty to develop one's talents. But these duties do not hold *against* the self because it would make no sense for one to claim or demand performance of some act one is unwilling to perform. Any claim-right presupposes some potential conflict of wills between the right-holder and some second party, but no such conflict of wills is possible within one and the same person. Hence, one's moral right to life cannot include any moral claim-right not to be killed by oneself. Nevertheless, the ground of the moral claim-right not to be killed by another person does seem to apply to the act of suicide. In many and perhaps most cases, suicide is not a victimless act, for the deceased has suffered a premature loss of life. More generally, under normal circumstances the act of killing oneself would cause grievous harm by depriving oneself of all the value one would have experienced were one's life to have continued. Also, (2) the act of killing oneself normally harms many of one's associates—family, friends, and colleagues. One's intimates will feel sorrow, grief, and often guilt. One will have deprived those who depend upon one in various ways, especially family and colleagues, of one's services. And all who enjoy one's companionship will have lost an important value in their lives. (3) It does not seem, however, that the act of killing oneself threatens the security of other persons in the agent's community in the way that other acts of homicide do. This reason is probably inapplicable to suicide. Therefore, there are in general, i.e. in typical cases, two reasons for one not to kill oneself: the fact that committing suicide would harm oneself and that it would harm one's associates.

But are these duty-imposing reasons? They are, in general, reasons for an agent not to kill herself, but are they also reasons for others to react negatively to anyone who attempts or does commit suicide? (1) The fact that one who commits suicide generally causes grievous harm to oneself may possibly be a reason for others to react negatively to anyone who is attempting to or has committed suicide, for it shows a destructiveness that reflects a gross disregard for the value of one's life. One who would thus destroy one's own life might well destroy or damage things of great value to others, their property, limbs, or even lives. One need not be confident that this line of reasoning is sound, but it does have some plausibility. (2) One should be convinced, however, that the fact that one who commits suicide generally causes harm to his or her associates is a reason for others in society with one who attempts or commits suicide to react negatively to that agent. It is dangerous to associate with anyone who thus disregards the welfare of others. Surely one has reason to disapprove of and even take preventive action against such persons. Thus, there exists at least one, and there might be two, moral duty-imposing reasons not to commit suicide.

It follows that there is no innocent moral liberty to commit suicide. One has an innocent moral liberty to perform some kind of action when there is no moral duty-imposing reason not to do so. But, we now see that there is at least one moral duty-imposing reason not to kill oneself. However, this is not the end of the story, for there might be a suspect moral liberty to commit suicide. One has a suspect liberty to perform some action when there is a duty-imposing reason not to do so, but this reason is overcome by some liberty-conferring reason. Two such reasons come easily to mind. One's duty not to commit suicide might be over-ridden by some more stringent duty. Some sacrificial suicides might be justified in this way. For example, an army officer who has been captured by the enemy and is facing torture probably has a moral liberty to kill himself by taking poison in order to avoid violating his duty not to reveal military secrets. In other situations, one's moral duty not to commit suicide might be undermined by excessive sacrifice. Ought implies can, not only in the sense that one has no moral duty to do the impossible, but also in that one is not morally bound to make heroic sacrifices. For example, although firefighters and others may have a duty to enter a burning building to rescue those in peril, no one has a moral obligation to do so when the building becomes an inferno and is about to collapse. Similarly, the moral duty not to commit suicide of a convict who, in a less humane era, has been sentenced to be repeatedly flayed until dead was probably undermined by the fact that refraining from killing herself would impose upon her a morally excessive sacrifice. To be sure, these are exceptional cases. It is difficult to believe that there is under normal or near-normal circumstances any liberty-conferring reason sufficient to overcome the general duty not to kill oneself. Thus, we need not affirm any general moral liberty, any liberty in normal circumstances, to commit suicide. What does emerge here, however, is that there are at least two exceptions to the general duty not to kill oneself. Therefore, there are at least two much more limited moral liberties to commit suicide, when killing oneself is the only way to avoid violating a more stringent moral duty or when to refrain from doing so would demand excessive sacrifice.

Each of these liberty-conferring reasons is applicable to some, only a very few, patients. For example, the moral duty not to commit suicide of a terminally ill patient who is receiving medically futile but very expensive care might be over-ridden by his moral duty not to exhaust the financial resources of his wife and children, especially if they understand and support his decision to end his life. And the moral duty not to commit suicide of a patient who is suffering intolerable, irremediable suffering can be undermined by the fact that to refrain from killing herself would impose a morally excessive sacrifice upon her. Hence, some patients do have a moral liberty to commit suicide.

There remains the question of whether these patients have a moral liberty to use assistance, such as lethal medication, provided by their attending physicians in exercising this right. Well, why not? Perhaps using assistance provided by one's physician to commit suicide would make one an accomplice in the moral wrongdoing of one's physician. There are those who argue that, whatever may be true of laypersons, it is inconsistent with the role of a physician or with the ethics of the medical profession for any physician to assist a patient with killing herself. Let us grant, merely for the sake of the argument, that physicians have a moral duty not to assist their patients to commit suicide, that by using assistance provided by one's physician a patient would become an accomplice in the physician's violation of this duty and that everyone has a moral duty not to become an accomplice in any violation of a moral duty. It would then follow that, even though some patient has a moral liberty to commit suicide, she would have a moral duty not to do so by using assistance provided by her physician. Nevertheless, this duty could be over-ridden by a stronger moral duty not to exhaust the financial resources of other members of one's family or undermined by the excessive sacrifice that would be imposed upon a patient suffering from intolerable, irremediable suffering. Thus, the very same liberty-conferring reasons that ground a moral liberty of some patients to commit suicide would also ground a moral liberty to use assistance provided by their physicians to do so. Therefore, although there is a general moral duty not to kill oneself, under exceptional circumstances some patients do have a moral liberty to use assistance provided by their physicians to commit suicide.

5.2. The Liberty to Request Assistance

The moral right to physician-assisted suicide, as earlier defined, is a rights-package consisting of three liberty-rights. The second of these is the right to request or not request assistance to commit suicide from one's attending physician. It might seem obvious that those patients who have a moral liberty to use assistance also have the less controversial moral liberty to request such assistance, but this could be challenged.

An opponent of physician-assisted suicide might argue that no patient has a moral liberty to request assistance to commit suicide from her physician because to do so would be to exploit her physician. To exploit someone is to make use of that person selfishly or unethically, to take unfair advantage of that person. Surely one has a moral duty not to exploit anyone, especially one's physician who has been providing and continues to give urgently needed medical care. But is requesting assistance from one's physician really exploiting her? It may well seem so, for it might be either asking too great a favor of one's physician as a friend or trying to impose an unprofessional obligation upon one's physician as a physician. Requesting one's physician to assist one to kill oneself seems to be imposing an excessive burden upon a trusted and loyal friend, for this may be asking her to act contrary to her conscience and will be in most states where it is a crime to assist another to commit suicide to ask her to expose herself to criminal penalties and/or the loss of her license to practice medicine. Alternatively, it is attempting to impose an unprofessional obligation to participate in killing upon one whose profession is defined by the goals of promoting health and preserving life. Plausible as this argument is, it is unconvincing. Merely to request assistance in committing suicide is not to exploit one's physician as long as it is understood that one's request imposes no moral obligation to provide the requested assistance and thus she is free to refuse. However, it probably would be exploiting one's physician to put pressure on her or attempt to coerce her into doing as one has requested.

To assert a moral liberty to perform some action is to deny the existence of any moral duty to refrain from doing so. But someone could plausibly argue that patients in fact do have a moral duty to refrain from requesting assistance to commit suicide from their physicians because to do so would be inciting their physicians to engage in immoral action. Plausible replies are that merely to request assistance falls short of inciting one's physician to provide assistance or that providing assistance to patients who intend to kill themselves in order to fulfill a stringent moral obligation or to escape intolerable, irremediable suffering is not immoral. However, let us grant the premises of this argument and reply as before. Even, if every moral agent does have a moral duty not to incite another to engage in immoral conduct and assisting anyone to commit suicide is morally wrong and requesting assistance is to incite, the implied duty not to request assistance can be overcome in the case of some patients either by the fact that suicide is necessary in order for them to fulfill some more stringent duty or because suicide is necessary in order for them to escape from intolerable, irremediable suffering.

An opponent of physician-assisted suicide might suggest that this reasoning is flawed. It never is necessary for a patient to request assistance from her physician in order to commit suicide because there are always others, family or friends, who could provide poisons as lethal as any prescribed medication or, if all else fails, a plastic bag with which one could suffocate oneself or a weapon with which one could kill oneself. This suggestion is sometimes unrealistic and always irrelevant. It is unrealistic because some, perhaps many, patients who intend to commit suicide will be unable to enlist the assistance of anyone other than their physician. More generally, the relevant question is not whether it is necessary to request assistance from one's physician, but whether it is necessary for the patient to request assistance from anyone. Notice that the premises we have accepted, merely for the sake of the argument, imply a general moral duty not to request assistance from anyone to commit suicide. Now, if this general duty is overcome by the moral liberty-conferring reasons that requesting assistance is necessary either for fulfilling a more stringent duty or for escaping intolerable, irremediable suffering, then patients in either of those situations do have a moral liberty of requesting assistance from someone. It would be incoherent to admit that a patient has a moral liberty to request assistance from someone but to deny that she has a moral liberty to request assistance from her physician on the ground that she could request it from someone else because, by this same reasoning, she would not have a moral liberty to request assistance from someone other than her physician on the ground that she could request assistance from her physician. In other words, what overcomes the general moral duty not to request assistance from anyone is the necessity of requesting assistance from someone. Thus, the moral liberty-conferring reasons that overcome this general moral duty are sufficient to ground a more limited specific moral liberty of some patients to request assistance from their physicians in order to kill themselves.

5.3. The Liberty to Obtain Assistance

The third component of the moral right to physician-assisted suicide as defined earlier is the liberty-right to obtain or not obtain assistance from one's physician that will be or at least could be used to commit suicide. Although such assistance, typically lethal medication and advice on how to use it, is most often provided to a patient who has requested it, a few physicians might offer it unasked. Do some patients really have a moral liberty to obtain assistance from their physicians, in most cases by accepting a prescription for lethal medication and having it filled, that could then be used to commit suicide?

It might be thought that this liberty is ruled out by their duty not to implicate their physicians in their contemplated suicide. Obtaining assistance involves anyone who provides it in a way that merely requesting assistance does not. Although a physician can avoid legal or moral responsibility for assisting a patient to kill herself by refusing her request for assistance, once assistance has been offered and accepted, the physician is necessarily implicated. Hence, by obtaining assistance from her physician, the patient is exposing her physician to the risk of criminal penalties or professional discipline that could extend even to the loss of his license to practice. Nevertheless, even granted that the patient does have a moral duty not to impose such burdens upon her physician, this duty would be overcome by the liberty-conferring reasons that it is necessary to accept assistance in order to fulfill a more stringent conflicting duty or in order to escape intolerable, irremediable suffering.

There is another way in which obtaining assistance goes beyond merely requesting it. Obtaining assistance that would enable a patient to commit suicide, whether requested or offered unasked, puts her in a position to commit suicide whenever she so chooses even when to do so would be morally wrong. It might seem that every moral agent has a moral duty never to put oneself in a position to violate a moral duty, but this cannot be true. When one makes a promise one puts oneself in a position to violate one's duty to do as one has promised, but surely one has no moral duty never to make a promise. Still, one may well have a general moral duty not to put oneself in a position where one will be sorely tempted to violate a moral duty. For example, someone addicted to gambling probably has a moral duty not to accompany a friend to a casino, no matter how firmly one has resolved not to gamble oneself, because this would put one in a position to be sorely tempted to violate one's moral duty not to gamble away the financial resources upon which one's family depends. Similarly, some patients might be sorely tempted to use the obtained assistance to commit suicide because of immoral family pressures or suffering that could be remedied even though they probably have a moral duty not to kill themselves under those conditions. However, those to whom the liberty-conferring reasons earlier specified apply, that it really is necessary to commit suicide either to perform a more stringent duty or to avoid intolerable, irremediable pain, would be acting under very different conditions, and would not have any moral duty not to kill themselves. Thus, the duty not to put oneself in a position where one would be sorely tempted to violate a moral duty, even if genuine, is inapplicable to those patients whose moral liberty to obtain assistance is defended here. Therefore, some, but not all, patients do have a moral liberty to obtain assistance from their physicians to commit suicide.

5.4. Protective Perimeters

How far have we progressed? We have defined a moral right to physician-assisted suicide as a rights-package consisting of the rights to request or not request, to obtain or not obtain, and to use or not use assistance from one's physician to commit suicide. Because each of these component rights is a liberty-right, the defining core of each is a moral liberty. We have seen that some patients do have the most controversial of these moral liberties. The liberties to refrain from requesting, obtaining, or using lethal medications are seldom challenged. But any real right is more than its defining core; it must also include associated positions that confer dominion over its core upon the right-holder. In the case of the rights to request, obtain, and use assistance, these must include at least what H. L. A. Hart called a protective perimeter—one or more duties of others not to interfere with, not to prevent or coercively hinder, the exercise of the core liberty.[5] Are the moral liberties of some patients to request, to obtain, and to use assistance protected by any such duties of noninterference?

There are several general moral duties that in various ways ground protective perimeters for these specific moral liberties. (1) *The duty not to obstruct dutiful action.* When it is necessary for a patient to commit suicide in order to perform some over-riding moral duty, then the duty not to obstruct dutiful action is applicable. This general duty implies the more specific duties not to interfere with the patient's liberties to request, obtain, and use assistance because under these circumstances to prevent or hinder the patient from exercising these liberties would be to obstruct dutiful action.

But why is there any moral duty not to obstruct dutiful action? Whatever moral reasons render some kind of action a duty for an agent also ground a duty of others not to prevent or hinder that agent in the performance of her duty. For example, making a promise imposes a moral duty upon the promisor to keep that promise. The fact that a moral agent has promised to do something is a reason for that agent to so act because, should he break his promise, the promisee will probably have relied upon him to her detriment and, in any event, his failure to keep his promise will weaken her trust in him. And it is a reason for those in society with the agent to react negatively to him if he fails to keep his promise because it shows him to be unreliable and untrustworthy, character traits damaging to most personal relationships. Moreover, the fact that an agent has promised to do something imposes a *prima facie* duty upon others not to prevent or hinder him in keeping his promise. This fact is a reason, although not always a conclusive reason, for others to refrain from forcing him to break his promise because were he to do so, the promisee who has relied upon him will probably suffer harm, at the very least inconvenience, and her trust in him may be undermined. And it is also a reason for those in society with anyone who does prevent or hinder the promisor from keeping his promise

[5] H. L. A. Hart, *Essays on Bentham* (Oxford: Clarendon Press, 1982), pp. 171–173.

to react negatively to the obstructer because it reveals him to be unconcerned with the welfare of others and destructive of that trust that is so important in personal relationships.

Similarly, one has a moral duty to rescue someone in distress, provided one is in a position to do so and doing so would not require excessive sacrifice. That some action would rescue another from distress is a reason for an agent to so act because failing to do so would cause preventable harm to the person in distress. And it is a reason for those in society with any agent who fails to rescue to react negatively toward her because this failure shows her to be malevolent or at least callous. Similarly, one who prevents or hinders an agent from rescuing someone in distress indirectly causes preventable harm to the person in distress and shows himself to be malevolent or callous.

This reasoning applies directly to a case of sacrificial suicide. A patient who commits suicide to prevent the financial ruin of or some other disaster for other members of her family will be performing her moral duty to rescue them from distress. Under some circumstances, this duty may outweigh her moral duty not to kill herself. Now the fact that her action will prevent great harm to her family is reason for her to commit suicide. And it is a reason for others to react negatively, although with understanding and sympathy, were she not to do so because her failure to rescue her family from financial ruin would show that she is either weak-willed or lacking in altruism. The fact that her suicide would rescue her family from distress is also a reason for others not to prevent or hinder her from requesting, obtaining, and using assistance from her physician when this is necessary for the performance of her duty to rescue because such obstruction would result in great harm to the members of her family. And it is a reason for those in society with anyone who does interfere with her exercise of these moral liberties to react negatively to the person who interferes because his interference reveals him to be malevolent or at least callous. Thus, the duty not to obstruct dutiful action is one ground of the protective perimeters around the liberties of some patient to request, obtain, and use assistance from their physicians to commit suicide.

(2) *The duty not to impede another's escape from harm.* One of our fundamental moral duties is the duty not to harm another person. The fact that some action would harm another person is a reason for a moral agent to refrain from performing that action. And it is also a reason for others in society with any moral agent who acts contrary to this moral reason to react negatively to that agent because by his act of harming he has shown himself to be malevolent or at least callous as well as dangerous to his associates, all sociability factors that seriously damage personal relationships. This very general moral duty implies various more specific duties, such as the duties not to injure another, not to damage or destroy their property, and not to impede their escape from harm. Thus, one has a moral duty not to push a swimmer who has jumped from a sinking ship away from a lifeboat, or not to close and lock a fire door with someone on the other side trying to escape from a burning room, or not to restrain a woman so that she cannot resist or flee from a rapist.

The same duty-imposing moral reason applies to the case of a patient seeking to commit suicide in order to escape from intolerable irremediable suffering. To interfere with her exercise of her liberties of requesting, obtaining, and using assistance from her physician when this is necessary in order for her to end her suffering is to harm her seriously by forcing her to endure continuing avoidable suffering. That impeding her escape from suffering would harm the patient is a strong, but not necessarily over-riding, reason for an agent not to so act. And it is a reason for those in society with anyone who interferes with the patient's liberties to react negatively to that agent because by so acting he has shown himself to be malevolent or at least callous and dangerously willing to inflict harm upon his associates. Therefore, the moral duty not to impede another's escape from harm implies protective perimeters around the core liberties of some patients' moral right to physician-assisted suicide.

(3) *The duty not to intrude into the life of another.* One literally intrudes when one invades someone's home without her permission, but what makes this morally wrong is the way in which it disturbs the person living there and disrupts her life. Wiretapping one's phone and reading one's correspondence are similarly intrusive. But the kind of intrusion that concerns us here is preventing or hindering one from acting on one's decisions. Such interference is intrusive because it disrupts one's activity. Choosing and acting, pursuing goals, and deciding how and when to do so are central to the life of any moral agent. To interfere with the freedom of action of another is thus to intrude into that person's life in a disruptive manner. Moreover, this intervention is correctly experienced as an intrusion into one's life by an external and hindering force, an alien will.

Accordingly, there are two morally relevant aspects of intrusive action. It disrupts the inner life and overt action of the person subjected to it, and it invades or forcefully enters that life from the outside. To the degree that it disrupts the life of another it is at least experienced as frustrating and often damages one or more of the projects of that person. As an invasion by some alien will, it is resented and alienates the person subjected to interference from the intruder. The harms that it imposes upon the person whose life is invaded—feelings of distress, disruption of her projects, and alienation from the invader—are reasons for an agent to refrain from intruding into the life of another. And those in society with an agent who intrudes into the life of another have reason to react negatively to him because by intruding he has shown himself to be selfish or at least self-centered and insensitive to the feelings and needs of others. Intrusive action is destructive of personal relationships because it tends to provoke retaliation if the person subjected to it is powerful, or withdrawal and isolation if she is relatively powerless. Therefore, there is a general moral duty not to intrude into the life of another.

However, this is only a *prima facie* duty and often over-ridden by some contrary moral reason, because trivial intrusions are easily justified and serious ones sometimes required by urgent moral considerations. But the duty not to intrude into the life of another by interfering with her action is seldom over-ridden when she is acting on a private decision, for this would be meddling. *The American Heritage Dictionary* defines the verb "to meddle" as "to intrude into other people's affairs or business; to interfere." Because any private decision of another moral agent is none of one's business and not one's concern, one's moral duty not to intrude into the life of that agent by interfering with her action almost always holds fast.

Meddling is only one way of intruding into the life of another, but an especially immoral species of intruding. Why is one's moral duty not to meddle so strong? The classic argument against interfering with the private action of another is that of John Stuart Mill.[6] His utilitarian defense of individual liberty correctly presupposes that what ultimately determines the morality of interference is its impact on the happiness, or more generally the well-being, of those affected by it. But his appeal to the nonmoral value of

[6] J. S. Mill, *On Liberty in Collected Works of John Stuart Mill*, J. M. Robson (ed.), Vol. 18 (Toronto: University of Toronto Press, 1977).

happiness cannot explain the moral duty not to meddle in the private action of another. The reasons that impose a moral duty must be specifically moral reasons—dual-aspect, essentially social reasons hinging on sociability factors, characteristics that determine the value of personal relationships. Moreover, his criterion for the identification of private decisions and actions, that they not harm anyone other than the individual agent, renders his reasoning inapplicable to any morally significant decision. A woman's decision to have an abortion may harm her husband who wants to become a father; a professor's decision to retire usually harms those graduate students still writing dissertations under his supervision; and someone's refusal to marry her lover may cause him to become clinically depressed. Even a trivial action such as my act of purchasing a candy bar harms someone else, for otherwise I would have deposited the 50 cents I spent in the Salvation Army kettle as I left the drugstore.

What definition of a private decision would be more useful for the purposes of moral theory? We may profitably conceive of the decision of an agent as private when no one else has a sufficient reason, a reason that outweighs any contrary reasons, to interfere with the agent's acting on it. Mill's conception is doubly mistaken. First, he assumed that a decision is private only if no one else has even the slightest reason to prevent or hinder the agent from acting on it. But, it is only the lack of a sufficient reason to interfere that renders a decision private. Second, he argued that the prevention of harm is the only kind of reason that could justify limiting the liberty of another individual. But a caregiver, such as a parent or guardian, may have a sufficient reason to prevent or hinder the person for whom she is responsible from acting in a manner that would make it harder or even impossible to carry out her duty of care. Presumably, in defining a private decision one should leave open the question of what kinds of reasons justify interference with the action of any moral agent.

By defining a private decision in terms of the lack of sufficient reasons for interference rather than in terms of harm to others, we make the statement that it is morally wrong to interfere with anyone acting on a private decision true by definition. Some might consider this result objectionable, for it seems to render what should be an important moral principle empty. It makes our concept of a private decision more like our concept of a murder, that by definition is an unjustified killing, than the merely descriptive concept of a homicide, that leaves open the possibility that the killing might be justified. Still, the definition we offer here is illuminating, for it indicates the reason why the duty not to meddle in the private action of another is an especially reliable species of the duty not to intrude into the life of another. The moral reasons that impose a general duty not to intrude are never over-ridden by any contrary reasons sufficient to justify meddling.

Any intrusion into the life of another normally disrupts that person's life to some extent and tends to damage her personal relationship with the intruder by provoking resentment. A person whose life has been invaded usually suspects that the intruder acted with ill will or at least treated her lightly—without careful consideration, without strong reason, and with lack of concern. Hence, she always could and sometimes will call upon the intruder to justify his treatment of her. If he can justify his action to her, if he can give her a reason sufficient to justify his intrusion into her life, then she will probably understand that his intrusion was not motivated by any malice or indifference to her well-being. Although she may still strongly dislike his treatment of her, she will resent it much less, if at all. Alternatively, if without actually challenging him she can imagine how he could justify his action to her, her resentment will be reduced or removed and the damage to their personal relationship repaired. But when the intruder has meddled in her life, then by definition he has acted without sufficient reason and is not in a position to justify his treatment of her. This renders the damage to their relationship irremediable, at least in this manner. Therefore, the fact that some intrusion would be meddling is a special reason for an agent not to intrude into the life of another in this manner.

It is also a reason for others to react negatively to any agent who does meddle, for by his action he has shown himself to be meddlesome. A meddlesome person is domineering; he imposes his will upon others arbitrarily or arrogantly. Because he interferes in the affairs of others arbitrarily, without sufficient reason, his intrusions tend to be unpredictable. We continually rely on others not to interfere as we go about our business or lead our everyday lives. But it is dangerous to rely upon meddlesome associates as we pursue our projects, for they may prevent or hinder the pursuit of our goals when we least expect interference and have taken no precautions to protect ourselves against their intrusions. In these ways, the fact that an agent has shown himself to be meddlesome is a special reason for others to react negatively to that agent. Therefore, the moral duty not to intrude is stronger when to intrude would also be to meddle in the private affairs of another.

Moreover, the duty not to intrude into the life of another is especially strong when to intrude would be to interfere with another intending to act on a momentous decision. This is primarily because the usual moral reasons that impose a duty not to intrude are magnified when a momentous rather than an ordinary decision is at stake. A momentous decision is one of utmost importance or significance, one having grave consequences. Accordingly, any intrusion into the life of an agent acting on a momentous decision will be especially disruptive, risk the gravest harm, and provoke the most intense resentment and complete alienation. But another consideration becomes relevant at this point. Most everyday decisions make very little, if any, difference to the meaning of one's life as a whole. Their significance, like their consequences, is limited and relatively unimportant. But whether or not one can carry out a momentous decision will change the meaning of a large part of one's biography and sometimes the significance of one's entire life. And the value of a life for the person living it depends much more on its positive or negative meaning or significance than on whether the various experiences that constitute it feel good or bad. This is why deciding for oneself whether to marry and, if so, with whom to share one's married life or determining one's own vocation or career in life are so very important. More generally, it is one reason why interfering with anyone acting on a similarly momentous decision is especially immoral. The disruption it causes in the life of another threatens the overall value of her life and the damage it does to her personal relationship with the intruder tends to be permanent rather then temporary.

These considerations apply to physician-assisted suicide. A patient's decision to commit suicide is clearly a momentous decision. It fundamentally affects her life for better or worse because, if acted upon, it ends her life and thereby terminates all her projects and eliminates all potential for future benefits or harms; at the same time, it profoundly affects the meaning of her life for herself and its significance for others. When committing suicide is necessary either to escape intolerable, irremediable suffering, or to perform a stringent duty to others, then the decision to do so is also a private decision. Although others may have some reason to interfere with the patient's execution of her decision to end her life, any reason others may have will not be sufficient to remove the excessive sacrifice the patient would suffer were she not to commit suicide or, because her decision will not as fundamentally affect the life of anyone else, over-ride the duty-imposing reason that imposes upon her a duty to sacrifice her life for others. We may presume, of course, that no individual act of physician-assisted suicide will have serious social consequences. (Whether introducing a legal right to physician-assisted suicide would be socially harmful is another question addressed in the previous essay.) Thus, others have a moral duty not to interfere with her exercise of her moral liberties of requesting, accepting, and using assistance from her physician when these are required to enable her to act on her private momentous decision to commit suicide. Therefore, the moral duty not to intrude into the life of another also implies protective perimeters around the core liberties in the patient's moral right to physician-assisted suicide.

(4) *The duty to respect the rational agency of others.* The *Oxford English Dictionary* defines respect as "deferential esteem felt or shown towards a person, thing, or quality,... the manifestation of a disposition to yield to the claims or wishes of another." Deferential esteem is not a mere feeling analogous to pain, for it involves an estimation or judgment that what is esteemed is worthy of deference. The object of the duty of respect, what is to be respected, is the rational agency of others, not their social status or achievements or even their moral virtue. And what is required by this duty is that one show or manifest this respect, or at least act as though one does respect the rational agency of others, by deferring to their choices and yielding to their actions. At the same time, it does not require one to agree that their decisions are rationally justified or their actions are morally right. Hence, it implies a general duty not to interfere with the actions of another rational agent even when one's own practical reason would dictate a contrary decision and different action.

The classic defense of the duty to respect the rational agency of others is that of Immanuel Kant.[7] Although his second formulation of the categorical imperative expresses a profound insight, we need not believe that his ethical theory can adequately explain either why one has this general duty of respect or what more specific duties are implied by it. He argues that rational agency commands our respect because the purely rational good will is the only thing that has unconditioned value. But the premise that its value is unconditioned implies nothing about its amount or degree of value. If its unconditioned value is minimal, then it deserves very little respect, hardly enough to justify more than a trivial *prima facie* duty. He also explains that our duty to respect the rational agency of persons is derived from our duty to respect the moral law. Because moral agency is one species of rational agency, this might somehow explain our duty to respect the moral agency of others. But, we also have a moral duty to respect the prudential agency of others, a duty that may be violated when one acts paternalistically. How this aspect of the duty to respect rational agency can be derived from the moral law remains obscure in Kant's moral theory. Moreover, because of his purely formal conception of rational agency, Kant's derivations of more specific duties of respect, such as the duties to develop one's talents and not to commit suicide, from the general duty to respect rational agency are unconvincing. However, impressive some of the recent versions of Kantian ethics may be, none of them seems to overcome these defects in Kant's original theory.

Can we do better? Can we explain the grounds of our moral duty to respect the rational agency of others? Well, a moral duty-imposing reason is both a reason for an agent to act or refrain from acting in some manner and for those in society with an agent

[7] I. Kant, *Groundwork of the Metaphysic of Morals*, H. J. Paton (trans.) (New York: Harper & Row, 1964).

who acts contrary to this reason to react negatively toward him or her. Why is the fact that interference would not respect the moral agency of another a reason for an agent not to interfere? First, one can have secure control over one's life only if others respect one's rational agency. A rational agent does not often act on the spur of the moment; she usually acts with a view to the future and with an eye on how each action will or will not fit with her past and future actions. Rational agency involves setting one's goals, selecting ways of achieving them, and integrating ends and means into a more or less coherent life. Thus, it requires one to adopt and carry out projects, often long-term and quite comprehensive projects. But this becomes impossible, or at least very difficult, if others do not respect one's rational agency enough to defer to one's decisions and yield to one's actions. It is by carrying out projects that one achieves one's most important goals, and it is these projects that give meaning and significance to one's life by making it more than a series of trivial satisfactions. Conversely, failing in those projects that define one's life makes one's life a failure. Therefore, to disrespect the rational agency of another by interfering with her action is to reduce her control over her life and thereby to threaten what matters most to the value of her life.

Second, one's self-respect is psychologically conditioned upon the respect or disrespect of others. Respect, in the sense that is relevant here, involves more than noninterference. One might refrain from interfering with the actions of another from indifference. Treating another with respect consists in deference to her decisions and yielding to her actions because one esteems her rational agency worthy of such treatment. And it is because respect involves a positive attitude that the respect of others affects one's self-esteem. Thus, self-respect consists in esteeming one's own rational agency worthy of deference by others. But if others fail to treat one with respect, one tends to lose confidence in one's ability to choose and act in a rationally justified manner. This loss of self-respect in turn diminishes one's initiative, creativity, endeavor, persistence, and self-reliance—all character traits that greatly affect the value of one's own life and the contribution one can make to the lives of those in society with one. These harmful consequences of not treating another with respect are reasons for any agent not to interfere with the actions of another rational agent.

Anyone who does interfere with the actions of another rational agent shows himself to be disrespectful. Why is the fact that someone is a disrespectful person a reason for those in society with that person to react negatively to him? A person who is disrespectful, in the relevant sense, tends to be arrogant, hypercritical, and domineering. These are all sociability factors that seriously damage personal relationships. Arrogance destroys the equal respect that is required for the highest forms of friendship and loving relationships. Hypercritical attitudes cause others to conceal their true feelings and beliefs, thus eliminating openness and leading to misunderstandings in one's relations with others. Those in society with a domineering person usually respond either by resisting his interference or by becoming passive, both incompatible with the cooperation needed for people to interact fruitfully with one another. Therefore, there are moral reasons that ground a moral duty to respect the rational agency of others by deferring to their decisions and yielding to their actions.

Now intolerable, irremediable pain can sometimes destroy or seriously interfere with a patient's rational agency, and a patient confronting the decision whether to kill herself or impose upon her loved ones some financial or other disaster can become so depressed that she cannot make a rational decision. But most attending physicians and consulting psychologists do not believe that patients who choose to commit suicide under normal conditions are always or even almost always acting irrationally. Hence, the duty to respect the rational agency of others sometimes implies protective duties of noninterference around the moral liberties of some patients to request, accept, and use assistance from their physicians to commit suicide.

What is established here, if the above line of argument is sound, is that there are protective perimeters of noninterference around the moral liberties of some patients to request, to obtain, and to use assistance from their physicians to commit suicide. When the exercise of these liberties is necessary to enable a patient to perform some stringent duty, then they are protected by the duty not to obstruct dutiful action. When their exercise is necessary to enable a patient to escape intolerable, irremediable suffering, then they are protected by the duty not to impede another's escape from harm. And in both situations, they are protected by the duty not to intrude into the life of another and by the duty to respect the rational agency of others.

Can this be correct? The four moral duties on which the conclusion here rests are all very general duties. Hence, they might seem to imply that it is always morally wrong to interfere with any attempted suicide. But surely one is sometimes permitted, indeed sometimes morally required, to prevent another from committing suicide. Although this is true, it is not incompatible with my conclusion. For one thing, these duties concern one's treatment of other rational agents. Interference may be permitted or even a duty when the person attempting suicide is not a fully rational agent or is not exercising her rational agency on this occasion. For example, it is permissible to confine a mentally ill and self-destructive person in an institution and to prevent her from killing herself while she is receiving treatment, or to prevent a drunken person who is temporarily despondent from throwing himself in front of a train. Again, these four duties are *prima facie* and may be over-ridden by some more stringent moral duty. For example, interference may be morally permissible if preventing a suicide or suicidal act is required by one's duty of care for that person. Thus, a parent or physician who has a duty to care for an adolescent and protect her from harm might under normal circumstances have a duty to prevent her from killing herself either deliberately or by engaging in some excessively risky conduct. Or one may have a duty to prevent another from an immoral act of suicide such as killing oneself to evade one's obligations to others or merely to get even with one's parents or a lover who has jilted one.

However, we have seen above that when committing suicide really is necessary either to enable a patient either to perform a very stringent duty or to escape from intolerable, irremediable suffering, the patient does have the moral liberties to request, to obtain, and to use assistance from her physician. It follows that under these circumstances suicide can be a fully rational and morally justified act. So, we may assert only that given these very special conditions and when the patient has and is exercising her rational agency, her moral liberties to request, to obtain, and to use assistance from her physician are protected by *prima facie* moral duties of others not to interfere.

5.5. Conclusion

The conclusion here is that there really is a moral right to physician-assisted suicide, more precisely a rights-package consisting of the moral liberty-rights to request or not request, to obtain or not obtain, and to use or not use assistance provided by one's attending physician to commit suicide. Each of these constituent rights includes at least a moral liberty that defines its essential content and *prima facie* moral duties of others not to interfere with the exercise of this core liberty. It very probably includes other associated positions, such as a moral immunity of the right-holder against having either her core liberty or any of its protective duties of noninterference extinguished by some unilateral act of another, but these complications have been omitted for the sake of brevity. The above reasoning shows only that two classes of patients possess this moral right to physician-assisted suicide: those who must commit suicide in order to fulfill a very stringent moral duty and those who can escape from intolerable, irremediable suffering only by killing themselves. Whether any other patients also possess a moral right to physician-assisted suicide remains undecided.

And whether attending physicians ever have any moral right to provide assistance to a patient who decides to commit suicide also remains an open question. The arguments presented above to defend a patient's moral right to physician-assisted suicide do not logically presuppose any moral liberty, much less any moral duty, of her attending physician to assist her with committing suicide. However, the full moral significance and the practical importance of her right will depend upon whether her attending physician does have some sort of a moral right to provide the assistance she has liberty-rights to request, to obtain, and to use. Therefore, the reasoning in this essay is only one part of a more comprehensive inquiry into the morality of physician-assisted suicide. It is an offering to my philosophical colleagues as a first step in the right direction.

CHAPTER 6

THE CONCEPT OF FETAL RIGHTS

Over the past two decades medical science has provided us with greatly increased knowledge of the sources of harm to unborn children. These include, but are not limited to, illnesses of the pregnant woman, side-effects of prescribed medication she is taking, environmental chemicals and radiation, the use of alcohol or tobacco or illegal drugs by the pregnant woman, and defects inherited from the genetic makeup of the biological parents. To a considerable but lesser extent, medical science now offers increased ability to prevent or reduce these harms to the fetus. Medicine can sometimes predict the risk of genetic or other harm in time to enable prospective parents to modify potentially harmful behavior, use new reproductive technologies to reduce the risk, or even avoid conception altogether. It can often diagnose genetic defects or medical problems of the fetus before birth and, in some cases, intervene either to abort a seriously defective fetus or to provide therapy *in utero*.

These developments have led some physicians, jurists, and moral philosophers to advocate the introduction of new fetal rights into our legal system. An almost-new example is the right not to be born with a life not worth living caused by medical malpractice. Although wrongful life suits claiming some such right are not new, almost none have survived at the appellate level. As a consequence of the increased ability to diagnose and treat medical problems of the unborn child, many argue that the fetus ought to have a legal right to medical care independent of but parallel to the mother's right to care by her physician. More novel than these extensions of medical malpractice law is the proposal that the fetus ought to be given analogous rights holding against her parents, especially her mother, not to be subject to parental malpractice and not to be denied any medical care necessary to preserve her life or promote her health.

The debate over these proposed fetal rights continues today and is unlikely to disappear in the foreseeable future. Those who advocate new fetal rights are vigorously opposed by those who believe that any such rights would infringe upon the fundamental rights of pregnant women and, in the bargain, do more to harm than to prevent injury to the unborn child. The moral and prudential questions raised by these proposed fetal rights are important, but I will postpone them for the next essay. Here I wish to identify and assess several of the conceptual problems concerning the very idea of fetal rights.

6.1. Conditional Rights of the Unborn

English common law has for centuries recognized the right of an unborn child to inherit property. John Salmond describes this fetal right as follows:

> There is nothing in law to prevent a man from owning property before he is born. His ownership is contingent, for he may never be born at all; but it is

nonetheless real and present ownership. A man may settle property on his wife and the children to be born of her. Or he may die intestate and his unborn child will inherit his estate.[1]

The traditional common law doctrine is that this is "present ownership," that is, a right now possessed by the unborn child, but that it is conditional on birth. If the child perishes in the womb, the inheritance will revert to someone else. This conditional right of the unborn has long been accepted in United States law.

More recently, the right of the fetus not to be wrongfully injured before birth has been recognized in our tort law. The leading case here is *Bonbrest* v. *Kotz*.[2] Prosser and Keeton describe the present legal situation thus:

> The child, if he is born alive, is now permitted in every jurisdiction to maintain an action for the consequences of prenatal injury, and if he dies of such injuries after birth an action will lie for his wrongful death.[3]

Once again, this legal right of the unborn child is conditional on birth.

These descriptions, generally accepted as authoritative, of two fragments of the law pose an obvious conceptual paradox. How can these rights be rights of the *unborn* child if they are conditional on *birth?* Well, when do these rights vest? If they do not vest until the child is born, then they cannot be rights of the unborn child. But if the unborn child has no right to inherit and no right not to be wrongfully injured, then presumably no right is violated when no inheritance is set aside for the unborn child or when someone wrongfully injures a fetus. This is clearly not the way our courts interpret the law. Still, if these rights vest before birth, why is it that the unborn child has no power to sue for a remedy until after birth? As the Supreme Court of Pennsylvania observed, it is a basic principle of tort law "that the cause of action accrues to the individual on the occurrence of injury causing damages."[4] Why has United States law made an exception to this general principle regarding these rights of the unborn?

6.2. Why the Born Alive Rule?

The fact that these rights of the unborn are conditional on birth is referred to as the "born alive rule." There is no logical or legal impossibility in abandoning this rule. The law could, if legislatures or courts so chose, confer unconditional rights upon the unborn child, rights of the fetus *qua* fetus. To be sure, this would require giving the fetus the legal power to institute legal proceedings before birth. Some might question whether this is possible. How could the fetus take legal action when it is not yet capable of acting in anything like the sense in which a normal adult human being can act? But as Justice Papadakos remarked:

---

[1] J. Salmond, *Jurisprudence*, 6th ed. (London: Sweet & Maxwell, 1920), p. 277.

[2] 65 F. Supp. 138 (1946).

[3] K. W. Page et al. (eds.), *Prosser and Keeton on the Law of Torts*, 5th ed. (Saint Paul MN: West Publishing Co., 1984), p. 368.

[4] *Amadio* v. *Levin*, 501 A.2d 1085 (Pa. 1985), at 1096.

> This inquiry cannot be directed to the physical capacity to comprehend and direct the filing of a complaint, for it cannot be doubted that there are numerous actions maintained by persons acting in a representative capacity for plaintiffs suffering legal disabilities (e.g. infancy, incompetence), or that these actions would survive the death of the disabled plaintiff.[5]

Similarly, the law could enable parents or other guardians to take legal action in the name of the unborn child.

Nevertheless, there may be reasons in the logic of the law, the legal implications of a fetal right not conditional on birth, that justify retaining the born alive rule. Any such reasons would probably vary from right to right. For example, courts might well reason that it would do nothing to advance and might sometimes frustrate the purpose of the law to recognize an unconditional right to inherit of the unborn child. This conditional right has at least two purposes. First, it respects the will of the deceased father to provide for all his children, a wish either expressed in the will he executed or presumed if he dies intestate. It accomplishes this goal by requiring that the appropriate portion of the father's estate be reserved for the unborn child should she be born; but if the child is not born, then she does not actually take possession of this property. This is as it should be, for a child who is never born needs no provisions to sustain her life. Second, the conditional right to inherit enables the father, whether or not he wishes to do so, to fulfill his legal duty of child support. It does this by giving the father the legal power to settle a share of his estate upon his unborn child. But an unconditional right to inherit would not achieve this aim any more fully, for if the child is never born, then the father's duty of child support does not apply to this child. Thus, abandoning the born alive rule would do nothing to advance the purposes of the legal right to inherit of the unborn child. In fact, it might under some circumstances even frustrate those purposes. Were the unborn child to take possession of the property before birth, then upon her death her inheritance would become her own estate. Since she would die intestate, her property would be distributed in the usual manner to any survivors, some of whom might not be children of her father, and thus might not either accurately reflect the will of her father to provide for his children or maximize his ability to fulfill his legal duty of child support. This justification for making this right of the unborn child conditional on birth is at least plausible and perhaps compelling.

There is an analogous argument for retaining the born alive rule for the right of the unborn child not to be wrongfully injured. Presumably the primary purpose of this legal right is to compensate the parents for the medical care of their wrongfully injured child, or to compensate the child for its medical expenses after the child is born and requires medical care beyond that received by her pregnant mother. This purpose fits admirably with the born alive rule. Were this rule abandoned, this right could also serve the purpose of compensating the parents for the wrongful death of their unborn child. But this is unnecessary because the parents already have an independent cause of action for wrongful death. Indeed, were the born alive rule eliminated, the parents might in some cases be awarded double remedy by suing in the name of their unborn child and at the same time suing in their own names. This double compensation would be undeserved by the parents and would impose an excessive burden on the defendant. My only reservation

[5] 5 *Amadio* at 1094.

with this argument is that there are exceptional cases, like *Amadio*, in which the parents lose their standing to take legal action for wrongful death so that they can receive no compensation for their loss unless their unborn child has a right not conditioned on birth not to be wrongfully injured.

Another reason the courts have refused to recognize any unconditioned right of the unborn not to be wrongly injured is that this would enable the parent or some other guardian to take legal action before the child is born and even when it perishes before birth. Any such suit would require the court to decide whether the evidence presented by the plaintiff has demonstrated that the act of the defendant was the proximate cause of the injury to the fetus and to assess damages in an amount appropriate to that injury. Both of these matters remain highly speculative while the fetus remains *in utero* but become much more manageable after the child is born. Although this consideration probably was very serious until recently, increased medical knowledge has greatly reduced its weight if not rendered it almost irrelevant.

A third argument against abandoning the born alive rule is that the unborn child could not be given any unconditioned and independent right not to be wrongfully injured because she is not yet an independent person. Until the child is born, she is biologically tied to and dependent upon her mother. Hence, a child cannot become a third party to any medical malpractice suit between a pregnant woman and her physician until the child is born and thus becomes a separate individual. One cannot assess this argument until one has answered another controversial conceptual question.

6.3. When a Separate Individual?

When does a child become a separate individual? Courts confront this question because they need to answer it in order to decide some legal issue. But an answer that provides a sound basis for deciding one legal issue might be irrelevant to a very different one. Hence, we must begin by distinguishing the various legal questions at issue in prenatal injury cases before we attempt to determine which answer is appropriate to each of them. (1) When is there a separate injury to the child? Anyone claiming compensation for medical malpractice due to negligence must establish four things: some injury or damage suffered by the plaintiff, that some act of the defendant was the proximate cause of this injury, that the defendant owed a duty of care to the plaintiff, and that the defendant breached his duty. Accordingly, any court that must decide a prenatal injury case in which the plaintiff is a child, rather than her parents, must determine when there is a separate injury to the child.

Until 1946, United States courts held that there is no separate injury to the child until the child is born. It grounded its decisions upon the dictum of Justice Holmes in *Dietrich* v. *Inhabitants of Northampton*[6] that a child *en ventre sa mere* is a part of its mother. Hence, any injury to the unborn child is simply an injury to a part of the mother, like an injury to her head or heart, and not a distinguishable injury to the child. This reasoning seems to fly in the face of contemporary medical knowledge. It is now often possible to diagnose physical injuries, genetic defects, and some diseases of the fetus *in*

---

[6] 138 Mass. 14 (1884).

utero. It might, nevertheless, be possible for courts to accept these facts and still reason that there is no need for the law to recognize these *as injuries* until the child is born because before birth the child does not suffer from them. For example, the child does not suffer from being crippled until it begins to walk and is not disadvantaged by brain damage until it begins to play with siblings or attend school. However, these postnatal disadvantages are surely the consequences of prenatal injuries, and the child could not claim compensation for them in and of themselves because the wrongful act was not their proximate cause. Also, there is some evidence that a fetus can suffer pain.

The case that first recognized the tort of prenatal injury, *Bonbrest* v. *Kotz*,[7] held that there is a separate injury to the child as soon as the child is viable. It rejected "... the assumption that a child *en ventre sa mere* has no juridical existence, and is so intimately united with its mother as to be a 'part' of her and as a consequences is not to be regarded as a separate, distinct, and individual entity."[8] As Justice McGuire opined:

> As to a viable child being "part" of its mother—this argument seems to me to be a contradiction in terms. True, it is in the womb, but it is capable now of extra-uterine life—and while dependent for its continued development on sustenance derived from its peculiar relationship to its mother, it is not a "part" of the mother in the sense of a constituent element—as that term is generally understood. Modern medicine is replete with cases of living children being taken from dead mothers.[9]

In short, the child becomes a separate individual when it is capable of being separated from its mother and surviving. Therefore, the child can suffer a separate injury as soon as it is viable. As far as it goes, this reasoning is surely sound. The unborn child is not merely a part of its mother, and viability is a sufficient condition for the capacity to suffer a separate injury. But is it also a necessary condition?

Other courts have ruled that there can be a separate injury as soon as the child is conceived. In *Zepeda* v. *Zepeda*,[10] Justice Dempsey reasoned as follows:

> The case at bar seems to be the natural result of the present course of the law permitting actions for physical injury ever closer to the moment of conception... How can the law distinguish the day to day development of life? If there is human life, proved by subsequent birth, then that human life has the same rights at the time of conception as it has at any time thereafter. There cannot be absolutes in the minute to minute progress of life from sperm and ovum to cell, to embryo to fetus, to child.[11]

This argument seems to presuppose that the fertilized ovum is a live genetically individuated human organism from the moment of conception and that it remains the same individual throughout its growth from cell to embryo to fetus to child born alive. Because its genetic code is different from that of its mother, it is a distinct individual and

---

[7] *Bonbrest* at 138.
[8] *Bonbrest* at 139.
[9] *Bonbrest* at 140.
[10] *Zepeda* v. *Zepeda*, 190 N.E.2d 849 (1963).
[11] *Zepeda* at 853.

can suffer a separate injury. And because it is a human life, it possesses all the rights that United States law confers upon human beings. The first conclusion seems sound, but the second could be challenged.

In *State* v. *Merrill*,[12] the defendant indicted for fetal homicide argued that the state criminal statute was unconstitutional because it was fatally vague.

> People will differ on whether life begins at conception or at viability. People may differ on whether death is the cessation of brain activity (an activity not present in an embryo) or the cessation of a functioning circulatory system. The problem, says defendant, is that absent statutory criteria, judges and juries will provide their own definitions which will differ, leaving the statutes vulnerable to arbitrary and discriminatory enforcement.[13]

The court rejected this reasoning as follows:

> The difficulty with this argument, however, is that the statutes do not raise the issue of when life as a *human person* begins or ends. The state must prove only that the implanted embryo or the fetus in the mother's womb was living, that it had life, and that it has life no longer. To have life, as that term is commonly understood, means to have the property of all living things to grow, to become.[14]

But is this the sense of the word "life" relevant to legal issues concerning the life, injury, or death of a child? The connection between a life, the biography of a human being, and biological life remains controversial. In any event, it seems entirely irrelevant to the question of whether an injury to the fetus is a separate injury from any injury to the mother.

(2) When does the physician owe a duty of care to the child? Showing that there was a separate injury to the fetus is only part of what an injured child must establish in any medical malpractice suit. She must also demonstrate that by causing this injury the physician breached his duty of care to her. But when is the child owed a separate duty of care by her mother's physician? United States courts long considered it impossible for the physician to owe any duty of care, a duty arising from the relation of physician to patient, to an unborn child. The special relationship between physician and patient traditionally was, and for most purposes still is, regarded as contractual. Surely the fetus is not capable of entering into any legally binding contract with the physician treating her mother. Hence, it would seem to be impossible for the physician to owe any contractual duty of care to the unborn child.

However, one party who enters into a contract with a second party may owe a contractual duty to some third party. For example, when a parent contracts with a babysitter to care for her child, the babysitter acquires both a duty to the parent to provide the agreed care and a duty of care to the child as a third-party beneficiary. And in *Zepeda* v. *Zepeda*, an illegitimate child sued his father for breaching his agreement to marry his mother. One of the theories upon which the plaintiff relied was that his father owed him a contractual duty as a third-party beneficiary.[15] Similarly, a physician who contracts to treat a patient

[12] 450 N.W.2d 318 (Minn. 1990).
[13] *Merrill* at 323.
[14] *Merrill* at 324.
[15] *Zepeda* at 852.

can sometimes thereby acquire a duty of care to some third party. The California Supreme Court decided that psychological therapists treating Prosenjit Poddar, a patient who had confided his intention to kill Tatiana Tarasoff and who subsequently did so, breached their duty to warn Tatiana of this threat to her life. The defendants argued that they had no special relationship to Tatiana that would impose upon them a duty of care to her, but the court held that her relation to their patient was sufficiently close to impose upon them a duty of care to her as a third party.

> Although plaintiffs' pleadings assert no special relation between Tatiana and defendant therapists, they establish as between Poddar and defendant therapists the special relation that arises between a patient and his doctor or psychotherapist. Such a relationship may support affirmative duties for the benefit of third persons.[16]

Presumably, analogous reasoning would establish a physician's duty of care to an unborn child arising from his special relationship to her mother.

The question then becomes "when does the fetus become a third party?" After *Bonbrest* rejected the view that the fetus is merely a part of the mother, courts were free to regard an unborn child as a third-party beneficiary of the contractual relationship between the mother and her physician as soon as it becomes capable of receiving a separate benefit or suffering a separate injury. To my mind, this reasoning is plausible, but it seems out of touch with modern medical practice.

If the physician owes a duty of care to the fetus as a third-party beneficiary of his contract with the mother, this is an indirect duty of care. But modern medicine makes it increasingly possible for the physician to treat the unborn child directly. For example, the physician can test the fetus for many genetic defects by amniocentesis and even perform operations upon the fetus *in utero.* This suggests that the unborn child is a separate patient to whom the physician owes a direct duty of care.

This alternative theory was advocated by Justice Handler, dissenting in part, in *Berman* v. *Allan*.[17]

> Their negligence consists of the failure to render proper advice to Mrs. Herman as an *expectant* mother. Indisputably in this relationship the doctors were caring for the unborn child as well as the mother; the duty they owed Mrs. Herman enveloped a duty to the unborn child. The breach of that duty affects both.[18]

Justice Handler insisted that the physicians owed a duty of care directly to the unborn child both because they were giving medical treatment to the child as well as to the mother and because the breach of their duty resulted in injuries to the child as well as injuries to the mother. Although this theory has not yet been widely accepted by the courts, it strikes me as appropriate to modern medicine. It implies that the physician owes a duty of care to the fetus as soon as his medical treatment directly affects her because she then becomes his patient in her own right.

---

[16] *Tarasoff* v. *Regents of University of California*, Sup., 131 Cal.Retr. 14 (1976) at 23.
[17] 404 A.2d 8 (1979).
[18] *Berman* at 20, italics in original.

(3) When does the mother owe a duty of care to her child? After a child is born, the parents owe a duty of care to their child. Recently some jurists have argued that a mother owes, or at least should owe, a duty of care to her unborn child. If so, it might be a violation of this duty for a pregnant woman to ingest toxic substances such as heroin or for her to refuse medical treatment needed by her fetus. When could a child become a separate legal person to whom her mother owes a duty of care?

One way to establish a mother's duty of care to her unborn child would be for the courts to recognize a tort of parental malpractice analogous to the existing tort of medical malpractice. Presiding Justice Jefferson, speaking for the Second District California Court of Appeal, suggested this legal development.

> The "wrongful-life" cause of action with which we are concerned is based upon negligently caused failure by someone under a duty to do so to inform the prospective parents of facts needed by them to make a conscious choice *not* to become parents. If a case arose where, despite the due care by the medical profession in transmitting the necessary warnings, parents made a conscious choice to proceed with a pregnancy, with full knowledge that a seriously impaired infant would be born, that conscious choice would provide an intervening act of proximate cause to preclude liability insofar as defendants other than the parents were concerned. Under such circumstances, we see no sound public policy which should protect those parents from being answerable for the pain, suffering and misery which they have wrought upon their offspring.[19]

Were this suggestion generally accepted by state courts, United States law would include a new tort of parental malpractice presupposing the pregnant woman's duty of care to her unborn child. Presumably, her fetus would be recognized as a separate second party to whom she owes this duty from its conception, just as is the case with a physician's duty of care to the unborn child.

But how could a mother have a duty of care to her child before it becomes viable, given the decision of the United States Supreme Court in *Roe* v. *Wade*?[20] To some jurists it seems inconsistent to maintain that a pregnant woman has a duty to protect her fetus from harm and to consent to unwanted medical care for its benefit when, during the first two trimesters, she has a constitutionally protected right to abort her fetus. Presumably, if she does not wrong her fetus by killing it, she could not wrong it by any lesser injury. But in a case concerning a fetus on the borderline of viability, the District of Columbia Court made a distinction that seems to apply throughout a woman's pregnancy. "Thus, as a matter of law, the right of a woman to an abortion is different and distinct from her obligations to the fetus once she has decided not to timely terminate her pregnancy."[21] The reasoning seems to be that when a woman aborts her fetus, no child will be born to suffer the consequences of her action; but when a mother injures her fetus so that it is born with serious medical problems, there will be a separate individual who will suffer from her negligent or willful action.

---

[19] *Curlander* v. *Bio-Science Laboratories*, App., 165 Cal. Rptr. 477 at 488, italics in original.
[20] 410U.S.113 (1973).
[21] *In re A.C.*, 533 A.2d 611 (1987) at 614.

A second way for the courts to establish a mother's duty of care to her unborn child would be for them to apply child abuse and/or child neglect statutes to acts of a pregnant woman that injure her fetus. Whether such acts really are crimes depends upon the correct interpretation of these statutes, and on that matter there is no consensus. In a unanimous decision concerning a woman who used cocaine during her pregnancy, the District Court of Appeal of Florida ruled that the Florida child abuse statute does not apply to an unborn child.

> From its legislative history, it is clear that the Legislature *considered* and *rejected* a specific statutory provision authorizing criminal penalties against mothers for delivering drug-afflicted children who received transfer of an illegal drug derivative metabolized by the mother's body, *in utero*.[22]

Presumably, if the Florida legislature had intended its law to apply to the fetus, it could have made its meaning explicit in the language of its child abuse statute.

However, there is often room for disagreement about the correct interpretation of a law. In a case concerning a pregnant woman who was arrested for using an illegal drug and who confessed that her addiction continued after her arrest, the legal issue was whether an unborn child is a "person" within the meaning of the Kentucky child abuse statute.[23] Justice Leibson, speaking for the majority, argued that it did not. He relied most heavily upon the Preamble to the Maternal Health Act of 1992 to determine the legislative intent of the Kentucky General Assembly.[24] He also argued that to interpret the statute otherwise would render it so vague as to violate the requirement of due notice and constitutional due process limits.

> The mother was a drug addict. But, for that matter, she could have been a pregnant alcoholic, causing fetal alcohol syndrome; or she could have been addicted to self-abuse by smoking, or by abusing prescription painkillers, or over-the-counter medicine; or for that matter she could have been addicted to downhill skiing or some other sport creating serious risk of prenatal injury, risk which the mother wantonly disregarded as a matter of self-indulgence. The defense asks where do we draw the line on self-abuse by a pregnant woman that wantonly exposes to risk her unborn baby?[25]

If no clear line can be drawn, then the court should not interpret the statute in a manner that would render it unconstitutional. But Justice Wintersheimer, with whom Justice Lambert joined, argued to the contrary that the unborn child is a person within the meaning of Kentucky's child abuse statute.

> Civil law has long recognized that an unborn child is a person. 42 Am.Jur.2d *Infants* § 2 states that biologically speaking, the life of a human being begins at the moment of conception in the mother's womb, and as a general rule of construction in the law, a legal personality is imputed to an unborn child for all purposes which would be beneficial to the infant after birth.[26]

---

[22] *State* v. *Gethers*, 585 So.2d 1140 (1991) at 1142, italics in original.
[23] *Commonwealth* v.*Welch*, Ky., 864 S.W.2d 280 (1993).
[24] *Welch* at 283–284.
[25] *Welch* at 283.
[26] *Welch* at 285, italics in original.

Adopting this accepted rule for interpreting the law, the court should hold that the term "person" includes an unborn child. If so, a pregnant woman has a duty of care to her fetus that is violated by any act of child abuse.

In a case concerning a woman who continued to take heroin during pregnancy, the Fourth District Court of Appeal of California advanced a novel argument against interpreting the California child abuse statute as imputing a duty of care to a pregnant woman.

> The language of section 273a(1) itself strongly suggests that the section was not intended to be applicable to prenatal conduct. In order to commit the offense defined by the statute, the offender must be a person "having the care or custody of [a] child."... This requirement presupposes the existence of a living child susceptible to care or custody.[27]

I find this argument unconvincing. Whether or not an unborn child is susceptible to custody, presumably a pregnant woman can care for her fetus. She can eat a nourishing diet to promote its growth, consent to medical treatment necessary for its health, and abstain from actions that would seriously injure it. In any event, child abuse statutes could impose upon a mother a duty of care for her fetus whenever the state legislature explicitly includes unborn children within its child abuse statute or the state courts interpret a child abuse statute as implicitly applicable to unborn children.

We have seen that when the courts recognize a child as a separate individual may vary depending upon the legal issue at stake in the case. The judicial reasoning advanced to justify the decision of a court will be based upon three sorts of considerations. The facts of the case, for example whether the unborn child is viable or can be directly treated by the physician, will be crucial. Obviously, these facts must be related to the body of norms, both legal rights and duties that are not at issue and state interests or public policies, in order to justify any legal conclusion. But frequently obscured by the complexity of the reasoning are conceptual or logical considerations. It is these that concern us here.

In judging when there is a separate injury to the child, a court considers a cluster of overlapping concepts and decides which of these do or should imply legal personhood. The basic dichotomy was defined by *Bonbrest* as deciding whether an unborn child is a "part" of the mother or a "separate, distinct, and individual" entity.[28] If a court believes that separateness is what counts, then it will probably opt for viability, the time when the fetus can be separated from the pregnant woman and survive. But if it believes that it is individuality that matters most, then it will decide that the fetus becomes a legal person at conception when it is brought into existence with its own genetic code. Other concepts that other courts took to be decisive were "human life"[29] and "a human person."[30] By selecting some concept or set of concepts as relevant, the court will determine which facts legally imply that the unborn child is to be recognized as a separate person.

In judging when the physician owes a duty of care to the child, the courts have deliberated about an additional conceptual question. How should the law conceive of the physician–patient relationship? Although courts usually interpret it to be a special sort of

---

27 *Reyes* v. *Superior Court, EtG.*, App., 141 Cal.Rptr. 912 (1977) at 913–914.

28 *Bonbrest* at 139.

29 *Zepeda* at 853.

30 *Merrill* at 324.

contractual relationship, modern medicine suggests that someone becomes or is capable of becoming a patient when the physician does or could treat that person directly. In fact, this way of conceiving of the relationship is not entirely new. If a physician negligently injures a stranger he finds unconscious and to whom he gives first aid, the stranger can sue the physician for medical malpractice. Presumably, if the physician was practicing medicine upon the stranger, who had not entered into any contract with that physician, then the stranger must have been his patient.

In judging when a pregnant woman can be guilty of abusing her unborn child, the conceptual question becomes how to interpret the word "child" in the child abuse statute of the state in which the case is being heard. Although most courts have decided that it refers only to children who have been born, others have decided that it covers unborn children as well. Unless the state legislature has explained its meaning clearly, the court must in effect supply its own definition of the word "child" in order to justify its conclusion regarding any alleged child abuse of a fetus.

Thus, we can see that the concept of fetal rights is much more complex than one might imagine. It involves more than relating the concept of a right to the concept of a human fetus. Whether the law does or should recognize any fetal right involves a cluster of other concepts that may or may not be relevant to legal personhood as well as defining the interpersonal relationships involved and the language of applicable laws.

6.4. Temporal Puzzles

Some jurists have argued, as have I,[31] that so-called rights of the unborn conditional on birth are really rights of the child when born. But this is not the way United States law conceives of them. The traditional common law doctrine of the right of the unborn child to inherit is that, although the child is not at liberty to use the property until after birth, the child actually possesses before birth any property willed to her. Similarly, although a child cannot sue for compensation for prenatal injury until she is born, she has the legal right not to be negligently injured as soon as she becomes a separate individual *in utero*. Why have the courts not conceived of these as rights of the born child? Probably because they have assumed that if the unborn child did not possess these rights before birth, then no right would have been violated were no inheritance set aside for her before her birth or were she injured before she was born. But is this assumption necessary?

Presiding Justice Dempsey, delivering the opinion of a First District Appellate Court of Illinois in *Zepeda* v. *Zepeda*, traced the development after *Bonbrest* of the child's right to sue for compensation for prenatal injury.[32] At first, the viability of the child became the criterion upon which recovery rested. More recently, suits were being sustained where the child was not viable when the injury occurred. He then reasoned as follows:

> How can the law distinguish the day to day development of life? If there is a human life, proved by subsequent birth, then that human life has the same rights at the time of conception as it has at any time thereafter.[33]

---

[31] *Real Rights* (New York & Oxford: Oxford University Press, 1995), pp. 137–145.
[32] *Zepeda* at 852–853.
[33] *Zepeda* at 853.

Thus, the court recognized that the unborn child possesses the right, conditional on birth, not to be negligently injured from the moment it first exists as a separate individual.

In the next paragraph, Dempsey discussed the possibility of continuing the development of this legal right even further back in time and posed a much more puzzling question:

> But what if the wrongful conduct takes place before conception? Can the defendant be held accountable if his act was completed before the plaintiff was conceived? Yes, for it is possible to incur, as Justice Holmes phrased it in the Dietrich case, "a conditional prospective liability in tort to one not yet in being."[34]

But how is this possible? How could the physician have any duty of care to the child before the child exists?

Presumably the physician's duty not to negligently injure the child is, as Wesley Newcomb Hohfeld explained, the jural correlative of the child's right not to be negligently injured by the physician.[35] Accordingly, one way to explain how the physician could have this duty even before the child exists would be to assume that the child has the logically correlative right not to be negligently injured before she exists. But can a child have any right before that child exists as a separate individual? Surely not. There are not children and their rights as there are children and their parents, where the parents of a child always exist before their child, or children and their toys, where the toys a child receives might have been manufactured long before the child was conceived. Although a legal person possesses rights, her rights are not objects or entities that exist independently of her, like her car or even her clothes. The relation between children and their rights is more like, although not entirely like, the relation between children and their ages or children and their personality traits. A child has no age before she exists and probably no personality until after she is born.

Why is it impossible for a child to possess any legal rights before she exists? The law confers any right upon some right-holder by virtue of some status. For example, I possess the right to vote in Saint Louis as an adult resident of the city of Saint Louis, and an unborn child has the right to inherit property from her father by virtue of being his child. Were I not a resident of Saint Louis, I would have no right to vote in that city; and were an unborn child not a child of the deceased, she would have no right to inherit any of his property. In short, a necessary condition for the possession of any legal right is having the status or qualifications specified in the legal norm or norms that define this right. Because a child has no qualities or attributes of any kind before she exists, she cannot have the specific qualifications necessary for any legally recognized status and, as a consequence, cannot possess any right before she exists. Therefore, one cannot explain how a physician could owe a duty of care to a child before she exists by assuming that before she exists the child has a right not to be injured.

The only other possible way to explain how a physician could owe a duty of care to a child before she exists would be to assume that a first party can have a duty to some second party before that second party has the correlative right. But can a duty to some

---

[34] *Zepeda* at 853.

[35] W. N. Hohfeld, *Fundamental Legal Conceptions* (New Haven: Yale University Press, 1919), p. 36.

child exist before that child exists as a right-holder? Presiding Justice Dempsey, who held that a child possesses rights from the moment of conception, assumed that it can.

> It makes no difference how much time elapses between a wrongful act and a resulting injury if there is a causal relation between them. Let us take the hypothetical case of an infant injured after birth by a defective household device. Suppose, before the child was conceived, a manufacturer negligently made a space heater and sold it to a retailer who retained it in his store. After the infant's birth his mother purchased the heater and used it in the room of her child who was burned because of its faulty preparation. Would there not be a right of action against the manufacturer despite the fact the negligence took place before the child was conceived?[36]

He clearly supposed that this rhetorical question would be answered in the affirmative by anyone knowledgeable in the law of torts.

But he failed to solve the central conceptual puzzle. The manufacturer's legal duty to the child and the child's right holding against the manufacturer are logically correlative. Hence, the manufacturer's duty implies the child's right, and vice versa. But how could the existence of the manufacturer's duty, a duty that existed and was violated before the child was even conceived, imply the existence of the child's right when neither the child nor her right exists? The answer is that the law normally defines rights and duties, including the status one must have to possess any right or duty, in general terms. For example, the law imposes a duty upon anyone who manufactures products not to produce them negligently so that they could injure anyone who uses or is exposed to them, and a remedial duty to compensate anyone injured by the violation of this duty. The same law or laws that impose these duties upon all manufacturers confers upon anyone who uses or is exposed to a manufactured product a right not to be injured by it because it was negligently produced and, if so injured, a right to be compensated for that injury by the producer. It is these general definitions that establish in the law the mutual logical implications between a specific duty and the correlative right. What establishes the particular relation between duty-bearer and right-holder in any given case is the fact that that this manufacturer negligently produced the product that injured this individual claimant. But the particular time when the product was manufactured or the particular time when the claimant was injured are legally irrelevant, and thus escape the logical correlativity, because these do not enter into the general legal definitions of the duty and the right. Therefore, it is logically possible for a duty to some child to exist before that child exists as a right-holder. But if this is so, what reason is there to ascribe legal rights to the fetus before birth rather than to the child after birth?

6.5. Possible Right-Holders?

Indeed, is it even conceptually coherent to ascribe rights to a fetus? This depends upon one's conception of a right. H. L. A. Hart argued that the essential function of legal

---

[36] *Zepeda* at 853.

rights is to respect the choice of the right-holder.[37] For example, the owner's right to paint her house green confers upon her the option of either painting it green or not painting it green as she chooses, an option protected by duties of others not to interfere with her choice. But if this is true, then only beings capable of choice could possibly possess any right. For this reason, he at one time concluded that we should "not extend to animals and babies whom it would be wrong to ill-treat the notion of a right to proper treatment."[38] Neil MacCormick rejected Hart's choice theory primarily because it could not explain the rights of very young children.[39] He argued that the essential function of rights is to protect the interests of their possessors.[40] Thus, the neonate's right to be cared for by her parents protects the infant's interests in health and well-being. If one adopts an interest theory of rights, then only entities capable of having interests are possible right-holders. For this reason, some of those who hold an interest theory of rights infer that nonhuman animals are capable of having rights but plants and the planet earth are not. A third alternative is a claim theory of rights. Joel Feinberg argued that what is distinctive and important about rights is that they give the possessor standing to claim performance of some correlative duty.[41] A paradigm example is the creditor's right to repayment that confers upon the creditor the power to sue for payment in the event that the debtor fails or refuses to repay the loan when due. Presumably, then, only beings capable of claiming, of demanding something as their due, could possibly be right-holders.

Although there is something to be said for each of these conceptions of a right, I believe that a legal right is best conceived of as a system of Hohfeldian legal positions that confer dominion upon the right-holder in face of some second party in some potential confrontation.[42] Thus, the defining core of the creditor's right to repayment is her legal claim against the debtor to be paid the agreed amount on or before the due date. The practical importance of this right consists in the way in which it allocates dominion over this core upon the creditor in the event that the debtor fails or refuses to pay his debt, for the creditor then has the power to claim payment in the courts. But there is much more to the creditor's right than her legal claim to repayment. She has the bilateral liberty to sue for repayment or not to sue as she chooses. She also has the power to cancel the debt and the bilateral liberty to exercise or not exercise this legal power. These bilateral legal liberties are, as Hart pointed out, protected by legal duties of others not to interfere with her choices. And her legal claim is protected by her legal immunity against having her claim extinguished by any one-sided act of another party. Together these associated legal positions confer dominion, freedom, and control, over the defining core claim upon the right-holder when confronted by any recalcitrant debtor. Other species of legal rights have a legal liberty, a legal power, a legal immunity, or even a legal liability as their defining cores, but in any real legal right there are always associated elements that allocate dominion over the core upon the party who possesses the right.

---

[37] *Essays on Bentham* (Oxford: Clarendon Press, 1982), pp. 162–193.
[38] "Are There Any Natural Rights?" *Philosophical Review*, 64 (1955), p. 181.
[39] "Children's Rights: A Test-Case for Theories of Right", *ARSP*, 62 (1976), pp. 305–316.
[40] Rights in legislation. In P. M. S. Hacker and J. Raz (eds.), *Law, Morality and Society* (Oxford: Clarendon Press, 1977), pp. 189–209.
[41] *Rights, Justice, and the Bounds of Liberty* (Princeton NJ: Princeton University Press, 1980), pp. 143–158.
[42] *A Theory of Rights* (Totowa NJ: Rowman & Allanheld, 1985), pp. 81–107.

My dominion theory of rights is, like H. L. A. Hart's choice theory, a will theory of rights, but it differs in two crucial ways. Hart explains the essential relevance of rights to freedom by placing a liberty at the center of every right, but on my dominion theory the defining core of a right could be a claim, a power, an immunity, or even a liability rather than a liberty. Also, Hart's choice theory requires that central to every right is one or more bilateral liberties, liberties either to do or not to do some specific action as the right-holder chooses. My dominion theory allows for unilateral liberty-rights, for example the right to vote in Australia where the law imposes a legal duty to vote so that one is not permitted to choose not to vote.

Why should one interpret any legal right according to my conception rather than adopting some alternative theory of the nature of a right? Well, each of the theories of rights I have mentioned, and most others, are intended to capture what distinguishes the concept of a right from all the other concepts needed to formulate the most complete and illuminating theory of law. I have argued elsewhere that my dominion theory of rights best identifies what is most distinctive and important about the way in which rights function in any legal system.[43] Now if one adopts my conception of a legal right, then one must infer that only an agent, a being with all the psychological capacities required for rational action, could possess any right. This is because it would be idle and misleading to ascribe dominion, freedom, and control, to any entity incapable of exercising the legal liberties and powers that give a right-holder dominion over the defining core of a genuine legal right. Since no fetus has yet acquired the capacity for even rudimentary rational action, it is conceptually impossible for unborn children to possess any legal rights.

It might seem that I have overlooked one important legal reality. Long before they have acquired the full capacity for rational action, children can and do exercise their legal rights through their agents, usually their parents or other guardians authorized to take legal action in their name. If the law ascribes this vicarious agency to neonates, why not to fetuses as well? As Justice Papadakos argued, speaking for the Supreme Court of Pennsylvania in a wrongful death case:

> Apart from the present dispute, this can have no other meaning than that, whether born or unborn, the child is an "independent life." The question then is whether this "independent life" could have instituted an action prior to death. This inquiry cannot be directed to the physical capacity to comprehend and direct the filing of a complaint, for it cannot be doubted that there are numerous actions maintained by persons acting in a representative capacity for plaintiffs suffering legal disabilities (e.g. infancy, incompetence), or that these actions would survive the death of the disabled plaintiff.[44]

I do not doubt, nor do I deny, the legal reality to which Justice Papadakos refers. Still, I suggest that it is very misleading to conceive of such legal actions as actions of infants acting through their agents because it is logically impossible for a being incapable of action to act through a representative. To be sure, a child can be represented by her parent or guardian who takes legal action in her name. But in such cases, the parent or guardian is representing the interests of the child, not her agency. No one acting in the name of a

---

[43] *Real Rights*, pp. 105–113, 132–136.
[44] *Amadio* at 1094.

child could possibly be representing the dominion, the freedom, and control, of an infant who does not yet possess the capacity to act freely and exercise control. Therefore, it is more accurate to describe this sort of legal actions as cases in which a parent or guardian is exercising, not the rights of an incompetent child, but his or her own rights *qua* the parent or guardian of the child. One can and should describe the full reality of children's rights without ascribing legal rights to the unborn or newly born child.

The patient's legal right not to suffer from medical malpractice is a relevant, unproblematic, and clear illustration of the allocation of dominion to a patient *vis-à-vis* her physician. Its defining core is the legal claim of the patient holding against her physician not to be injured by his medical malpractice. Correlative to this is the physician's duty not to treat the patient without her consent and, if she consents, to exercise due care in his treatment of her. Notice that the patient has the legal liberty to consent or to refuse medical treatment as she freely chooses, but that the physician has no choice regarding whether his treatment of the patient is or is not legally permissible. In the event that the physician injures the patient by his negligent action, the patient has the power to sue for compensation and the bilateral liberty either to exercise or not exercise this power as she chooses. On the other hand, the physician has no legal power to cancel his duty of care to his patient or to extinguish her power to take legal action against him should he breach his duty to her. What better example could there be of the way in which a legal right confers dominion upon the right-holder in face of some second party?

To describe recent developments in United States law as conferring a legal right not to suffer from medical malpractice on the unborn child is seriously misleading. It suggests that the unborn child now has the legal power to consent to or refuse medical treatment and the legal liberties to exercise this power as she chooses. It also insinuates that the fetus could, were it injured by medical malpractice, take legal action or refrain from taking legal action to obtain compensation for its injuries. More generally, it suggests that the fetus can act to exercise its rights in all, or at least most, of the ways that a normal adult right-holder can. But this is impossible because an unborn child lacks any capacity for rational action and therefore cannot act in any of the ways that give meaning and substance to legal rights. Therefore, anyone who wishes to understand how the law really functions should deny that the human fetus is a possible right-holder.

6.6. Fetal Agency as Legal Fiction

What seems to make it impossible for the law to ascribe rights to the human fetus is the incoherence between the most accurate and illuminating conception of a right and the nature of a fetus. But this often puts a court between a rock and a hard place. In *Bonbrest* v. *Kotz*, for example, the court confronted the question of whether an infant has a right of action springing from a prenatal injury through professional malpractice with resultant consequences of a detrimental character. It would appear that after birth a child could not sue for a remedy because the physician's negligent action is not the proximate cause of the harms she is now suffering as a subsequent consequence of the prenatal injury. And if the fetus has no legal right not to be injured by medical malpractice, then tort law provides no remedy at all for the suffering caused by the wrongful action of the physician. But it is a basic principle of justice that the law ought to provide a remedy for every wrong, and surely a child who must throughout her life suffer the detrimental consequences of medical malpractice has been wronged. As Justice McGuire observed, quoting the Supreme Court of Canada:

> If a child *after birth* (italics supplied) has no right of action for prenatal injuries, we have a wrong inflicted for which there is no remedy, for, although the father may be entitled to compensation for the loss he has incurred and the mother for what she has suffered, yet there is a residuum of injury for which compensation cannot be had save at the suit of the child.[45]

To avoid this violation of natural justice, the court assumed that the unborn child is a person capable of action and, thus, capable of possessing legal rights.

This assumption is, of course, false. In other words, it is a legal fiction. Ian R. Kerr provides a helpful description of the nature of legal fictions.

> Generally, a legal fiction is a false assumption of fact made by a court, as the basis for resolving a legal issue.... Legal fictions, it is said, provide a mechanism for preserving the established rule while ensuring a just outcome.[46]

This surely explains why various courts confronting hard cases involving inheritance, wrongful death, or medical malpractice have falsely assumed that the unborn child has the psychological capacities required for the possession of legal rights. And if the only, or at least the best, way to reach just decisions in these cases is to make this false assumption, then the fiction of fetal agency is probably justified.

If the assumption that the fetus has the psychological capacities required for the possession of any real right is false, then it would seem that any statement by a lawyer or judge purporting to describe the rights of unborn children must also be false. But this is not so. The law is a text-centered practice.[47] It is a mistake to believe that one can know what the law really is merely by reading the published constitution, statutes, and judicial opinions. To discover the real law of any society one must look to the complex ways in which it actually functions in the practices of the legal institutions, especially the courts, of its legal system. But these practices consist in the creation and application of legal texts—often a written constitution, always statutes, and usually judicial opinions. Because many authoritative legal sources of United States law ascribe rights to the unborn child and because these texts are generally accepted and applied in our courts, fetuses really do have legal rights in our legal system. Therefore, statements by lawyers and judges ascribing legal rights to the unborn child are true.

Nevertheless, I insist that human fetuses are not possible right-holders. According to the conception of rights that best describes how they really function in a legal system, only a being capable of rational action could possess a right. Even a jurist who rejects my dominion model of legal rights should recognize that fetal agency is a legal fiction, for the law pretends that the actions of a parent or guardian representing an unborn child *are* the actions of that fetus. Presumably, then, any conclusions about the legal rights of the fetus grounded on the false assumption that the fetus is capable of rational action must

[45] *Bonbrest* at 141.
[46] *The Philosophy of Law: An Encyclopedia* (New York & London: Garland Publishing Inc., 1999), p. 300.
[47] C. Wellman, *An Approach to Rights* (Dordrecht: Kluwer Academic Publishers, 1997), pp. 174–175, 231–232.

themselves be false. Therefore, statements ascribing legal rights to the unborn child are false.

Have I not just contradicted myself? Not at all because I am not a lawyer; I am a philosopher of law. There are at least two perspectives from which one can speak about the law, roughly what H. L. A. Hart called the internal and the external points of view.[48] Lawyers and judges, when they are acting in their roles within a legal system, speak the language of the law. And in the authoritative language of United States law, the unborn child does have certain rights. But philosophers and jurists, who are acting as observers of the law and attempting to describe it in a theory that clearly and accurately explains how the legal system really functions, need to speak a different language. And in their theoretical language, the unborn child should not be said to have any legal rights.

Hart's distinction between the internal and the external points of view employs two metaphors, the spatial point from which one views an object or landscape and the spatial difference between the inside and the outside of a box or building. My reference above to two different languages is also a figure of speech. Although the human rights language of international law uses two languages, French and English, the only official legal language of the United States is English. Still, the language of the law does not wear its meaning on its face. Just as judges must interpret the statutes or judicial opinions they apply in any given case, so must anyone else attempting to understand any legal text interpret its meaning. And the appropriate interpretation depends upon one's purpose. The purpose of a judge's interpretation is to determine what meaning a text ought to have when applied within a legal system. This legal meaning consists in the legal implications of the text when applied to particular cases. The purpose of a philosopher's interpretation is to describe and explain how the text actually functions in a legal system. Because any illuminating explanation needs to be systematic, it must reveal differences in the ways that texts expressed in similar language actually allocate freedom and control under the law. Thus, a legal fiction can be both legally true and factually false because it can both have real legal consequences when accepted as true by a practicing judge and be known to be false by a theoretical jurist who is an observer of the world and of the law. Hence, judges should interpret ascriptions of rights to the unborn child in authoritative legal texts as referring to real fetal rights, but the conceptual analysis of the expression "rights of the unborn" for the purposes of legal theory need not and should not assume that the fetuses possess any legal rights. The rights of the unborn child recognized by our courts are real rights, but they are not really what they are said to be. They are not legal rights possessed by the fetus; they are rights actually possessed by others in their capacity *qua* parents or guardians. And when a parent or guardian exercises any of these rights, she is acting as a custodian rather than as an agent representing her unborn child.

6.7. Conclusion

We have seen that the concept of fetal rights is problematic in three distinguishable ways. First, its value in any explanatory theory of the law is in doubt. One of the goals

[48] *The Concept of Law* (Oxford: Clarendon Press, 1961), pp. 88–91.

of jurisprudence is to provide a conceptual analysis of the language of the law that will enable us to understand the nature of law in general and the logical connections between the various concepts used in any legal system. Philosophers disagree about the best way to interpret legal rights for the purposes of constructing an explanatory theory of the law. My dominion theory of rights, the view that the function of any legal right is to confer freedom and control upon its possessor, implies that only a being capable of agency could possibly possess any right. If one accepts my conception of legal rights and assumes, as empirical science establishes, that the human fetus is incapable of action in the relevant sense, then one must conclude that the language of fetal rights is conceptually incoherent. Those who hold most other versions of a will theory of rights must accept the same conclusion. Of course, this conclusion is controversial because philosophers and jurists disagree about which conception of a right will enable one to formulate the most accurate and illuminating legal theory. I have argued in this chapter that although the concept of fetal rights may be useful in the law, it is misleading in legal theory because it falsely suggests that fetuses can and do exercise their rights in the same ways that normal adult right-holders do.

Second, the concept of fetal rights is legally problematic because its logical presuppositions are often in doubt. Judicial reasoning that could justify any decision in favor of a child injured by medical malpractice performed before she was conceived must assume either that the plaintiff possessed even before she existed the right not to be thus injured or that the physician had a duty of care to that child even before she possessed the logically correlative right. Although both of these assumptions are conceptually puzzling, I have argued that the latter is intelligible while the former is not. More generally, any judge who would find in favor of a child alleging wrongful prenatal injury must assume that the child is taking legal action on the basis of her legal right. But how can an unborn child or even a neonate take legal action when she has not yet acquired the psychological capacities required for genuine human agency? Judges have resolved this conceptual problem by accepting the legal fiction of fetal agency and the analogous fiction of neonatal agency. What is at stake in resolving these conceptual issues is not choosing a conceptual analysis of the language of rights that enables one to formulate the best explanation of how rights function in the law but finding the premises a judge needs to justify a decision that most usefully and justly applies existing legal sources to the facts of the case before a court. Hence, the appropriate criteria for dealing with these conceptual problems in the concept of fetal rights are practical rather than theoretical.

Third, the concept of fetal rights is also legally problematic because its legal implications are often in doubt. If one grants that an unborn child has a right not to be injured by medical malpractice, then it would seem that even while the child is *in utero* a parent or guardian could take legal action against their physician in the name of the child. But we have seen that this would confront any court with the difficulties of determining before birth the existence and extent of prenatal injuries to the child, the proximate cause of any such injuries, and the fair amount of compensation for them. To avoid speculative and perhaps unjust resolution of such cases, without denying just compensation in the end, the common law and most subsequent United States legislation have adopted the conception of rights of the unborn child conditional on birth. Similarly, we have seen that when judges determine the time at which an unborn child becomes a separate individual, they are deciding when a prenatal injury counts as an injury for which a child may subsequently sue for compensation or when a physician owes a duty of care to the child or when a mother first owes a duty of care to her child. These judges are attempting to interpret the concept of fetal rights so that its legal consequences will result in just decisions when the law of fetal rights is applied to particular cases that are and will be before the courts. Here also the appropriate criteria are the practical norms of public welfare and social justice rather than the theoretical ideal of explanatory power.

Whether utility and justice require the introduction of new rights of the unborn child holding against the mother is today hotly debated in the United States. When jurists, moral philosophers, judges, and legislators attempt to resolve this controversy, they should look primarily to the urgency of the relevant public policies and to the demands of justice. But they will also confront deep and difficult logical and legal problems posed by the concept of fetal rights as it is applied in the law and employed in legal theory. This essay attempts to clarify some of these conceptual issues, leaving the normative issues for the next essay. I have discussed only how these conceptual issues arise in United States law because my knowledge of comparative law is very limited. But similar puzzles about the concept of fetal rights are implicit in the legal systems of at least Canada, England, Ireland, Australia, Switzerland, Austria, and Japan that have also recognized rights of the unborn child. Moreover, the same advances in medical science that now enable greatly increased prediction, diagnosis, prevention, and treatment of prenatal injury or defect will soon produce demands for new fetal rights around the world. Hence, the conceptual analysis of the language of fetal rights will be an important factor in legal developments far beyond the borders of my own country.

CHAPTER 7

MATERNAL DUTIES AND FETAL RIGHTS

Recent developments in medical science and medical technology have occasioned a number of proposals for extending the scope of existing legal rights of the unborn and the legal recognition of new fetal rights. The first article to receive widespread attention was "The Juridical Status of the Fetus: A Proposal for Legal Protection of the Unborn" by Patricia A. King.[1] She posed the central question as follows:

> What claims to protection can be asserted by a human fetus? That question, familiar to philosophy and religion, has long haunted law as well. While the philosophical and theological issues remain unresolved, and are perhaps unresolvable, I believe that we can no longer avoid some resolution of the legal status of the fetus. The potential benefits of fetal research, the ability to fertilize the human ovum in a laboratory dish, and the increasing awareness that a mother's activities during pregnancy may affect the health of her offspring create pressing policy issues that raise possible conflicts among fetuses, mothers and researchers.[2]

It is the conflicts between human fetuses and their mothers that soon became and remain today the focus of the most heated debates in legal periodicals, popular media, state legislatures, and courts.[3]

King noted that both adults and children have constitutional rights. But children are not yet capable of making rational decisions. Why, then, do we give them rights?

> We do so to increase the likelihood that they will be regarded as persons rather than property. Giving children rights also makes it easier for the state to protect them from the harmful acts of parents or third parties. Long ago, the Supreme Court said, "It is in the interest of youth itself, and of the whole community, that children be both safeguarded from abuses and given opportunities for growth into free and independent well-developed men and citizens." Ultimately, it may be this last trait that truly motivates courts and legislatures to give children rights—the potential to grow into mature, competent, well-developed adults.[4]

---

[1] Patricia A. King, "The Juridical Status of the Fetus: A Proposal for Legal Protection of the Unborn", *Michigan Law Review* 77 (1979), 1647–1687.

[2] *Id.* at 1647.

[3] But the terminology of "maternal—fetal conflict" probably oversimplifies the situation in most cases. See Cheryl M. Plambeck, "Divided Loyalties: Legal and Bioethical Considerations of Physician-Pregnant Patient Confidentiality and Prenatal Drug Abuse", *Journal of Legal Medicine* 23 (2002), 4–5.

[4] King, *supra* note 1m at 1668.

She then argued that the unborn are like children in their potentiality to become rational adults. Therefore, we have the same reason to confer legal rights upon the human fetus as we have to recognize the rights of neonates or young children. Although there are differences between a viable fetus and a newborn child, none are of the sorts that are relevant to awarding legal protection.[5]

Accordingly King proposed the legal recognition of new and expanded fetal rights. The most controversial of these are rights that would impose legal duties upon their mothers. She argued that the interests of a mother and her fetus should be weighed equally in resolving conflicts between them. This may require a pregnant woman to submit to treatments that she finds objectionable, such as blood transfusions, where they are necessary to save the life of her unborn child. And where the mother's personal activities, such as smoking, drinking, or using medication, endanger her fetus, it would be justifiable to intervene to prevent injury to her fetus.[6]

Another influential journal article was "Conditional Prospective Rights of the Fetus" by Margery W. Shaw.[7] After reviewing the findings of medical science regarding risks to the human fetus and the kinds of fetal therapy now offered by modern medicine, she described what she considered to be the maternal duties during early, middle, and late pregnancy. She summed up one of her controversial conclusions briefly:

> Thus, the stage was set to allow children to sue their parents for prenatal harms. In most cases, the tortfeasor would be the mother because she has direct control over her fetus during pregnancy. Negligent exposure to noxious chemicals and drugs, refusal to accept genetic counseling and prenatal diagnosis, refusal to obtain prenatal therapy, or failure to provide a modified diet, could give rise to a cause of action. It will take courage to reverse the well-established legal presumption that the mother's rights transcend those of the fetus. This presumption should hold only if the fetus does not become a living child.[8]

In one respect, this proposal is conservative. It retains the traditional doctrine that the rights of the unborn child should be conditional on birth. But it is radical both in the range of fetal rights it advocates and in the way that it would impose burdensome legal duties upon pregnant women.

In a series of articles[9] and more recently in *Children of Choice*,[10] John A. Robertson has advanced the most systematic case for the recognition of new fetal rights. Although he defended an expansive interpretation of the procreative rights of women and takes their right to bodily integrity seriously, he argued that the moral and legal rights of a pregnant woman are sometimes limited by the rights of her unborn child.

---

[5] *Id.* at 1669.
[6] *Id.* at 1683–1684.
[7] Margery W. Shaw, "Conditional Prospective Rights of the Fetus", *Journal of Legal Medicine* 5 (1984), 63–116.
[8] *Id.* at 95.
[9] See especially John A. Robertson, "The Right to Procreate and in Utero Fetal Therapy", *The Journal of Legal Medicine* 3 (1982), 333–366; and "Procreative Liberty and the Control of Conception, Pregnancy, and Childbirth", *Virginia Law Review* 69 (1983), 405–464.
[10] John A. Robertson, *Children of Choice* (1994).

> The mother has, if she conceives and chooses not to abort, a legal and moral duty to bring the child into the world as healthy as is reasonably possible. She has a duty to avoid actions or omissions that will damage the fetus and child, just as she has a duty to protect the child's welfare once it is born until she transfers this duty to another. In terms of fetal rights, a fetus has no right to be conceived—or, once conceived, to be carried to viability. But once the mother decides not to terminate the pregnancy, the viable fetus acquires rights to have the mother conduct her life in ways that will not injure it.[11]

In addition to the right not to be harmed by its mother, the fetus has a right that the pregnant woman submit to medical treatment necessary to protect the life and health of her unborn child. After discussing the permissibility of using experimental fetal therapies, Robertson insisted:

> A more likely source of conflict with the fetus' mother would result from the mother's refusal of a fetal therapy established as safe and effective. Such a situation arises from time to time when a mother refuses an exchange transfusion for a fetus suffering from Rh incompatibility. The conflict could also arise if in utero surgery for hydrocephalus or bilateral hyronephrosis, now in the experimental stage, became accepted procedures. A mother's refusal of therapy in these situations could be the basis for a civil suit or criminal prosecution if it resulted in death or injury to the fetus, just as a parent's refusal of necessary medical care for a child can now be the basis for civil or criminal liability.[12]

More recently, Robertson has reaffirmed his conclusion that these maternal duties and the correlative fetal rights are morally and legally justified as necessary to prevent prenatal harm to offspring.[13]

The logic of these proposals is simple. (1) They assume, usually with little or no argument, that pregnant women have a moral duty not to cause avoidable harm to their unborn children and a moral duty to submit to fetal therapies needed to preserve the health of children to whom they will give birth. (2) They also assume, on the basis of our knowledge of medical science and modern medical practice, that pregnant women will sometimes violate these moral duties with serious consequences for their offspring. Hence, they conclude that the law should recognize new fetal rights in order to enforce these maternal duties and thereby prevent avoidable harms suffered by the children who would otherwise be born with serious medical problems and be unable to develop into fully functioning citizens of our society. Let us examine each step in this reasoning.

7.1. Maternal Duties

Do pregnant women have a moral duty not to act in ways that would injure their fetuses and a moral duty to submit to fetal therapies needed to preserve the health of any

[11] Robertson, *Procreative Liberty and the Control of Conception, Pregnancy, and Childbirth*, 405, 438.
[12] *Id.* at 444.
[13] Robertson, *Children of Choice* 179–180.

children to whom they will give birth? The relation of a pregnant woman to her fetus is typically described in morals and in the law as the relation of mother to unborn child. This suggests that any duty of a pregnant woman regarding her fetus is a special case of, or at least analogous to, a duty of a mother regarding her child after birth. Hence, I will begin by examining the grounds of a mother's moral duties regarding her baby or young child and then consider how, if at all, they apply to a pregnant woman. To simplify my discussion, I will examine only the case of a mother, living in our society today, who has had sexual intercourse with her husband, conceived and carried her fetus to term, and delivered a normal healthy baby. However, the general principles to which I appeal should apply, mutatis mutandis, to all mothers in our society and to mothers in other societies at other times.

In this paradigm case, the mother has both a moral duty to care for her child and a moral duty not to abuse her child. Although her husband also has parallel moral duties to care for and not to abuse their child, it would be a mistake to reduce their respective duties to some set of shared parental duties. For one thing, the content of maternal duties may not be identical to the content of paternal duties. For another, the mother's duties remain even if her husband dies or abandons the family.

What, then, are the grounds of these maternal duties? The mother's moral duty to care for her child is grounded on her causal responsibility for the existence of her child as a dependent creature with basic human needs. By her actions of engaging in sexual intercourse, gestating her fetus, refraining from abortion and delivering her child, she has brought into existence someone who needs food, shelter, clothing, medical care, education, affection, and much more. Needs impose a more urgent moral imperative than mere wants, for needs are things such that if one does not have them, one suffers harm. Because young children are incapable of satisfying their own needs, they will be harmed if no one cares for them. And because the mother, with the assistance of the father, caused this needy dependent child to be born, she would be the proximate cause of any harm it would suffer were its needs not met. Thus, the mother's special moral duty to care for her child is grounded on her general moral duty not to harm another together with her special causal responsibility for creating a dependent being with unmet needs.

The logic of this moral reasoning probably requires further explication. Harm is a moral duty-imposing reason. This is to say, the fact that some action would cause harm to another person is a reason both for any moral agent not to perform that action and for those in society with any agent who acts contrary to this morally relevant consideration to react negatively to him. Hence, everyone has a moral duty not to harm others. One can harm another either by doing something that injures her, such as battering one's child, or by failing to do something that would have prevented harm to that person, such as rescuing a small child from the attacks of an older bully. Hence, the general moral duty not to harm another implies a general duty to rescue another from the threat of harm.

Accordingly, one who is in a position to rescue someone threatened with serious harm has a moral duty to do so, unless this would impose excessive sacrifice upon oneself. This general moral duty of any and every moral agent implies a special and more stringent special moral duty when one is causally responsible for exposing another to any serious threat of harm. Consider the example of a business enterprise that knowingly sends an employee into a foreign country where rebel or impoverished groups often kidnap employees of wealthy corporations. Unfortunately, some group known for killing unransomed captives does kidnap their employee and demands one million dollars for his release.

Presumably, anyone who has one million dollars to spare has some moral obligation to rescue the victim from this threat to his very life, but the business enterprise has a special and more stringent moral obligation to do so because they are the ones who exposed him to this peril. On a more personal level, consider a variation on an event in my own life. Imagine that I have invited a friend to accompany me on a camping trip to Yellowstone National Park, where I have rented a cabin. During our first night there, we hear from the safety of our cabin black bears overturning garbage cans to find food. The next morning, I suggest that my friend should collect firewood while I chop it up near our cabin. Having reduced the first batch of wood to fireplace size, I look up to see two huge bears about 15 feet from me and between my friend and our cabin. Presumably, anyone who is in a position to warn the endangered wood-gatherer and lead him to safety has a moral duty to rescue, but I have a special and especially stringent duty to do so because it is I who am causally responsible for his danger. Similarly, a mother has a special moral duty to care for her child because she is, in large part, causally responsible for the existence of the situation in which a human being unable to meet his own needs will suffer serious harm where she not to care for him.

She is not morally required to meet these needs all by herself. One would hope that her husband will do his fair share in caring for their child, and ideally other members of the family will participate in caring for the child to a greater or lesser extent. Also, our society offers various forms of assistance in meeting the needs of her child such as police protection, stores where she can buy food and clothes, medical clinics and hospitals, and even welfare benefits for dependant children. Still, upon the mother falls the moral duty of satisfying any unmet needs to the extent that she is capable of doing so without undue sacrifice to her own interests.

Because this maternal responsibility is so broad, given the range of basic needs of any child, and the ways in which these needs can be met are so various, given the great diversity of the circumstances in which mothers and their children interact, it is difficult if not impossible to define the content of the mother's duty of care with any precision in positive terms. Thus, the mother's duty of care can best be defined negatively as the moral duty not to neglect her child. She neglects her child only when she causes or threatens serious harm to her child by failing to provide what the child needs. This negative definition is also more illuminating, for her moral duty of care is grounded on the harm her child will suffer if its needs are *not* met. Nevertheless, this duty of care is primarily a positive rather than a negative duty. It requires for the most part that the mother perform those actions, such as feeding and clothing her child, that will meet its needs rather than simply refraining from acting in ways that will injure her child.

Legal duties are artificial rather than natural because they are typically imposed upon duty-bearers by the actions of legislators or judges and grounded upon statutes or court holdings. But the duty of care that I have been discussing is natural in the sense that it is not created by social institutions or imposed upon the mother by the actions of others. It is grounded on the moral relevance of a natural fact, the fact that neglecting the needs of one's child will, under normal circumstances, cause harm to that child. Nevertheless, this duty is conditioned by social institutions in at least three ways. First, the needs of the child, especially the kind of education and protection it needs in order to develop into an adult capable of living her own life, depends upon the institutions of a society, for example, its high-tech economy or its violent slums. Second, many of the institutions of a society providc ways of fulfilling the maternal duty of care. Public or private schools supplement the educational activities of the mother, an extended family or day care facilities enable a mother to hold a job to earn the means of satisfying a wide range of needs, and a program of food stamps or welfare for dependent children would do the same were she unable to support her family. Third, social institutions, such as arrangements for adoption or foster homes, might possibly even relieve her of her duty of care, her duty to satisfy the *unmet* needs of her child. If others are willing and able to meet those needs, nothing more is morally required of the mother.

In addition to a moral duty to care for her child, the mother in our paradigm example has a moral duty not to abuse her child. To abuse, in the relevant sense, is willfully or negligently to maltreat or injure, especially repeatedly. Presumably, everyone has a moral duty not to abuse any child grounded on the duty-imposing reason that to do so harms the victim. But the mother's special duty not to abuse her child is distinct from that general duty. It is her duty not to maltreat or injure her child physically or psychologically or to expose her child to the serious risk of any such injury. Examples of child abuse would be beating a child who fails to eat all his vegetables, harassing one's child verbally, or leaving one's child in the care of a known child batterer or sexual molester. This maternal moral duty is grounded on her special moral duty to care for her child. Because one of the basic needs of any child is for protection, a need greater than that of any normal adult who had learned to cope with his environment, the mother's duty of care implies the duty to protect her child. Abusing her child is an especially egregious violation of this moral duty because a child is especially vulnerable to his caregivers. Thus, the mother's negative duty *not* to abuse her child is derived from her positive moral duty to care for her child. Both are special duties because both depend upon her special procreative relationship to her child. And both are natural duties because both are grounded on the fact that child neglect or child abuse are harmful together with the causal responsibility of the mother for the birth of a vulnerable creature with basic needs.

However, the mother in our paradigm example also has moral duties regarding her child that are derived from the institutions of our society. The traditional family is a social institution defined by such roles as husband and wife, brother and sister, and mother and child. The *Oxford English Dictionary* defines a role as "an actor's part in a play, film, etc.; *fig*. the part played or assumed by a person in society, life, etc." Social scientists have given literal meaning to what was originally a figure of speech. Most of them identify a role with the requirements of a social status or the set of rights and duties attached to a social position. Accordingly, the social role of motherhood imposes specific institutional duties upon the mother of any child in our society. The content of the duties attached to any role is defined by the specific patterns of behavior expected of anyone who occupies the role. In our society, these include the mother's duty to care for her child and her duty not to abuse her child. Because the scope of these institutional duties is determined by the expectations of most members of our society, their content may differ somewhat from the mother's natural duties to care for her child and not to abuse her child. No doubt, these social norms are undergoing considerable change in our society today. Nevertheless, they still apply to the mother in our paradigm example.

Granted that the social role of motherhood imposes on the mother in our paradigm example the institutional duties to care for her child and not to abuse her child, are these duties of her social role also moral duties? Presumably, the hit man does not have any moral obligation to kill someone fingered by the mob even though this is his primary role responsibility. And many would argue that a public executioner ought morally to resign from his office rather than carry out the duties of his social role. Still, unless the actions required by a social role are immoral, one incurs a *prima facie* moral obligation to fulfill the duties of any role one accepts. This is because by accepting or assuming a social role one invites others in one's society to rely upon one to fulfill those duties. The duties of any social role are defined by the expectations of typical members of the society. That is to say, others expect anyone occupying that role to act in certain specific ways. They expect this in the descriptive sense that they believe that one will act in those ways, and they expect this in the normative sense that they demand it of one. Accordingly, they respond to one who has assumed any role in the light of these expectations. By accepting or assuming a role, one has invited others to rely upon one. If, contrary to their expectations, one fails or refuses to act as expected, others may have acted to their detriment and one will have betrayed their trust. Therefore, the duties of a mother's role are moral as well as institutional duties.

A clearer illustration of incurring a moral obligation by inviting others to rely upon one is the act of promising. When I promise a student that I will meet her in my office this afternoon to help her understand an assigned journal article, I invite her to rely upon me to be there at the agreed time and to do my best to enable her to interpret its obscure language. If she trusts me, she will probably postpone other tempting activities and refrain from asking other members of the class for assistance. If I fail to do as I have promised, then she will be poorly prepared for the mid-term examination later this week and will have relied upon me to her detriment. It is the harm I will have caused her that is the moral duty-imposing reason that grounds my moral obligation to keep my promise. It may not be accurate to say that the mother in our paradigm example has promised, even tacitly, to fulfill the duties of the social role of motherhood. Still, the moral obligation not to invite others to rely upon one to their detriment is more general and more fundamental than the moral duty to keep one's promises. Imagine that I am standing beside a river during a spring flood watching in fascination as bits of debris fly past me. Then, I notice a girl clinging to a log floating toward me. Fortunately, I have a coil of rope with me. I throw the rope to the girl, who gratefully abandons her log and grasps the rope. Surely, I have a moral obligation to hold fast to my end of the rope and pull her to safety. Were I to let go of the rope, she would have relied upon me to her detriment and, unless she could grab another piece of wood, I would have caused her grievous harm.

The mother in our paradigm example has voluntarily assumed the role of motherhood by engaging in sexual intercourse with her husband, gestating her fetus, choosing not to have an abortion and delivering her child. By her actions, she has invited other members of her family and society at large to rely upon her to fulfill her maternal duties to care for her child and not to abuse her child. An unmarried woman who has acted in the same ways presumably has the same moral duties, as does a woman who adopts a child and thus acquires the social role of caregiver. Whether a woman who has been raped or who has been prevented from aborting has any moral duty to carry out the role duties of motherhood is dubious. But I will ignore these complications in order to discuss a more relevant borderline case.

How, if at all, do these grounds of paradigm examples of maternal duties apply to a pregnant woman? Does she similarly have a moral duty to care for her fetus and a moral duty not to abuse her fetus? A pregnant woman is very often described as the "mother" or her "unborn child." This suggests that a pregnant woman assumes the social role of motherhood even before her child is born. But it is hard to know whether this language is best interpreted literally or figuratively. It is also possible that in our society there is a social role of pregnant woman defined by some sort of duties to care for and not to abuse one's fetus. Whether either of these possibilities is actual a mere philosopher is not qualified to judge. Only systematic empirical investigations of social scientists could establish this. However, it is clear that the expectations of most members of our society today are not determinate enough to define role duties that would imply that a pregnant woman is morally required to refrain from smoking, drinking, or jogging or to submit to fetal therapy against her will. No social consensus has yet emerged regarding these previously unimagined moral issues posed by new medical science and modern medical technology. Therefore, we can and should for present purposes ignore any maternal duties imposed upon a pregnant woman by her social role as expectant mother or pregnant woman.

On the other hand, we cannot disregard the moral grounds of the mother's natural duties to care for and not to abuse her child. These are her causal responsibility for the birth of a dependent vulnerable child with unmet needs together with the fact that her failure to care for her child or her actions of child abuse would cause serious harm. Do these same moral reasons ground analogous moral duties of the pregnant woman to provide for the basic needs of her fetus and not to injure her fetus or expose it to the serious risk of such injury? The answer depends upon whether fetal neglect or fetal abuse harms the fetus in the morally relevant sense. No doubt there is some sense in which a fetus really needs nourishment and medical care and can be injured by the pregnant woman's actions of smoking or drinking heavily. But whether one should consider this harm in the morally relevant sense depends upon the most adequate theory of value. My theory of value that locates all intrinsic value in conscious experience suggests that this is not so. No doubt insects need nourishment and are injured by being squashed. But I doubt that either capturing a firefly and neglecting to feed it or swatting a fly cause harm in any sense that would ground a moral duty not to do so. Probably, when a fetus has developed sufficiently to feel pain, one has a *prima facie* duty not to cause it to suffer. But aside from that, I believe that one should not take the language of maternal duties to care for and not to abuse one's fetus literally because fetal neglect or fetal abuse do not cause harm in the morally relevant sense. In any event, those who have advocated new fetal rights have not rested their case on any serious harms to human fetuses.

This does not imply, however, that a pregnant woman has no moral duties regarding her fetus. Although whether a pregnant woman's actions or omissions really harm her fetus is controversial, the harms her child might suffer after he is born are all too real and often very serious. A pregnant woman may be and sometimes is causally responsible for postnatal harms such as those resulting from her failure to provide adequate nourishment for her fetus or her use of prohibited drugs during her pregnancy. Thus, the grounds of a mother's material duties are clearly relevant to her unborn child. Therefore, a mother's natural moral duties to care for her child and not to abuse her child, although conditional on birth, impose moral constraints upon her behavior long before the birth of her child.

These duties are not, of course, unlimited. Their scope is restricted by morally relevant factors other than the harms their violation would cause. The mother's moral duty to care for her child, before or after birth, is limited by her ability to meet his needs without excessive sacrifice to herself. And only willful or negligent injury or exposure to injury constitutes child abuse. Nevertheless, these moral duties of a pregnant woman are very real and important enough to be taken most seriously by every future mother and by society in general. This confirms the assumption of those who propose new fetal rights that pregnant women have a moral duty not to cause avoidable harm to their unborn children and a moral duty to submit to fetal therapies needed to preserve the health of children to whom they will give birth.

7.2. Maternal Violations

They are also correct to assume that pregnant women will sometimes violate these moral duties with serious consequences for their offspring. Consider first the maternal duty not to abuse one's child, not willfully or negligently to act in a way that will cause it serious harm or expose it to the risk of such harm. There have been a number of notorious court cases[14] concerning mothers who have continued to use illegal drugs throughout their pregnancies in spite of warnings of the threat to their offspring. The harms that can be caused by fetal drug syndrome are now thoroughly established by medical science. Narcotic addicted fetuses are very often born prematurely and underweight. They are typically irritable, writhing, and sometimes convulsing. They suffer from vomiting, diarrhea, dehydration, and fever. Because neither alcohol nor tobacco is a prohibited substance, women are not similarly taken to court for heavy drinking or smoking, but modern medicine has also established that these common activities are also injurious to the unborn child. For example, fetal alcohol syndrome consists of growth retardation, facial anomalies, mental retardation, and assorted congenital defects affecting other organs. Late in pregnancy, immoderate exercise or sexual intercourse, both activities in which pregnant women sometimes engage even against their doctor's orders, can cause serious harm to the fetus and thus impose suffering upon the child after birth.

The case of *Stallman v. Youngquist*[15] concerns a pregnant woman who endangered her unborn child, subsequently born with serious injuries resulting from an automobile accident caused by her negligent driving. An earlier case[16] concerns a child born with deformities of his legs and feet caused by the negligent driving of his mother while she was pregnant. Presumably, any pregnant woman who exposes herself to high levels of radiation thereby exposes her unborn child to a serious risk of genetic defects. A woman who during the first trimester of her pregnancy associates with anyone infected by German measles might easily become infected herself with the result that her child may be severely retarded, blind, deaf, and suffer major malformations such as congenital heart defects. Although such actions would constitute a violation of one's moral duty not to abuse one's unborn child only if done negligently or willfully, it is highly probable that some pregnant women do yield to the temptation to risk the health of their future offspring.

There is also every reason to believe that pregnant women will at least as often violate their moral duty to care for their unborn child; that is, they will neglect to meet its basic needs including its need for medical care. Some women will persist in dieting so rigidly that their fetuses will suffer from malnutrition. Substandard nutrition during

---

[14] See especially *Reyes v. Superior Court, Etc.*, App., 141 Cal.Rptr. 912 (1977); *Matter of Baby X*, Mich.App, 293 N.W.2d 736 (1980); *State v. Gethers*, 585 So.2d 1140 (Fla.App. 4 Dist. 1991); and *Ferguson v. City of Charleston, S.C.*, 186 F.3rd 469 (4th Cir. 1999).

[15] *Stallman v. Youngquist*, 129 Ill.App. 3d 859 (1984).

[16] *Smith v. Brennan*, 157 A.2d 497 (1960).

pregnancy may result in low birth weight, which causes higher incidence of defects such as impairment of brain development and even mortality. Whether Jessie Mae Jefferson had any moral obligation to submit to a caesarean section and blood transfusion in spite of her religious scruples is debatable. In any event, the record shows that she refused her consent to this medical treatment that her physicians judged necessary for the health of her unborn child.[17] Not all women who refuse to submit to cesarean sections claim to be exercising their right to religious freedom. Some refuse to submit to an invasive medical procedure that they regard as too risky, and it may be that in some cases they have no duty to submit because the risk would constitute an excessive sacrifice for them. But blood transfusions are less intrusive and much less likely to cause harm. Still, some women would probably refuse to submit to a blood transfusion to their unborn child even though there is an Rh incompatibility between mother and fetus that, unless treated promptly, will destroy the blood cells of their unborn children. Because Christian Scientists often refuse to consent to medical treatment urgently needed by their young children, it is probable that those who are pregnant will even more often refuse consent to medical treatment needed for their fetuses. More recently, new medical technology has made it possible to perform surgery on the unborn child *in utero*, for example, to correct obstructive hydrocephalus. If this build up of spinal fluid inside the brain is not relieved by a shunt or other operation, it often results in infections, vascular problems and brain damage. Quite possibly, some mothers will insist upon postponing therapy until after their child is born even though this may be too late to prevent serious injury to their offspring.

There is even the case of Pamela Rae Stewart who seems to have violated both her moral duty to care for her unborn child and her moral duty not to abuse that child.

> She was warned by her doctor that hers was a problem pregnancy and was instructed to seek medical attention at the first signs of bleeding. She also was told to refrain from sex with her husband. But when her bleeding started, she disregarded the medical advice, stayed home with her husband, had sexual intercourse, took amphetamines, and waited twelve hours before going to the hospital. Her son was born with extreme brain damage and died six weeks later....[18]

How many women there are who would act so irresponsibly is hard to know. But we do know of actual cases in which some pregnant women have failed to care for and have even abused their unborn children. And there are strong reasons to believe that this has happened and will continue to happen in many instances of which the public will never become aware.

The fact that pregnant women will sometimes, and perhaps often, violate their maternal moral duties to care for and not to abuse their unborn children poses a serious social problem. The most obvious cause for concern is the suffering of the victims, the children who will be born with very serious injuries. These defective babies also impose additional burdens upon other members of the families into which they are born and sometimes disrupt the family unity. In addition, there are the costs imposed upon society. These are

---

[17] *Jefferson v. Griffin Spalding Cty. Hospital*, Ga. 274 S.E.2d 457 (1981).

[18] Martha A. Field, "Controlling the Woman to Protect the Fetus", *Law, Medicine and Health Care* 17 (1989), 118.

not only the financial expense and use of scarce medical resources required to treat children born needing exceptionally high levels of medical care, but also often the birth of persons who will be lifelong burdens upon their society rather than productive citizens.

7.3. Proposed Fetal Rights[19]

The most obvious solution to this social problem posed by pregnant women who neglect or abuse their fetuses would be to impose legal sanctions upon those who violate their moral duties to care for and not to abuse their unborn children. Thus, Patricia A. King entitled her influential article "The Juridical Status of the Fetus: A Proposal for Legal Protection of the Unborn." She proposed both criminal and civil sanctions, but for both philosophical and pragmatic reasons I shall limit my discussion to the latter. I agree with H. L. A. Hart that although the criminal law imposes duties, it does not confer rights in the strict sense. And attempts to protect fetuses by the use of criminal sanctions are likely to be counterproductive by causing pregnant women either to mislead their physicians or to neglect medical care entirely.[20] Other jurists, most notably Margery W. Shaw[21] and John A. Robertson,[22] have also proposed introducing into United States law new fetal rights holding against the mother. And Presiding Justice Jefferson, delivering the opinion of the California Court of Appeal in *Curlander v. Bio-Science Laboratories*, suggested that in the future parental malpractice concerning the unborn could be the basis for wrongful life suits just as medical malpractice now is.[23] No one to my knowledge has explained in any detail precisely what form these proposed fetal rights should take, but we need some sense of the available options in order to assess these proposals in any useful manner.

First, which fetal rights would be relevant? If the problem is that some, perhaps many, pregnant women will fail to fulfill their duty to care for their unborn children and might even abuse them, then the solution would seem to be to confer the logically correlative legal rights upon human fetuses. Accordingly, reformers might propose the introduction into our legal system of fetal rights to all needed care and not to be abused holding against their mothers. The very broad scope of these rights would be both an advantage and a disadvantage. It would be an advantage because the law would be general enough to cover the wide variety of ways in which irresponsible women might act in ways that would cause their offspring avoidable suffering. On the other hand, they might generate considerable numbers of trivial lawsuits thus unduly burdening our courts and, even worse, unreasonably constrain the freedom of pregnant women. Still, no right is absolute, and courts would no doubt limit these fetal rights in the light of the mother's constitutional rights to liberty and privacy and especially the parental right to exercise reasonable discretion in raising one's children.

Less likely to engender frivolous lawsuits would be more limited fetal rights to specific kinds of care and not to be abused in specified ways. For example, fetuses might

---

[19] See Carl Wellman, "The Concept of Fetal Rights", *Law and Philosophy* 21 (2002), 65–93.

[20] For this reason, I will not discuss cases like *Ferguson v. Charleston*, 532 U.S. 67 (2001) in any detail.

[21] Shaw, *Supra* note 6.

[22] Robertson, *Supra* notes 8 and 9.

[23] *Curlander v. Bio-Science Laboratories*, App., 165 Cal.Rptr. 477 (1980) at 488.

be given a legal right that the mother provide for all necessary medical care and a legal right that they not be abused by the use of prohibited drugs or excessive consumption of alcohol by the pregnant woman. If medical science shows that the greatest threat to the health and well-being of the children who will be born is posed by the lack of prenatal medical care and these forms of abuse, then these legal rights would go a long way toward solving the social problem while avoiding the disadvantages of much broader rights. And if such more specific rights were defined somewhat differently in various states, then experience would show how best to solve the problem posed by irresponsible pregnant women.

The reasoning of the court in *Smith v. Brennan* rests on the assumption that every child has the legal right to begin life with a sound mind and body.[24] If some such right were recognized more widely in the future than it has been to date, then this would provide legal remedies for both the pregnant woman's failure to provide care needed for the future physical and mental health of her fetus and any serious abuse of her unborn child. However, any such right seems to possess to an unacceptably high degree the disadvantages of very general fetal rights. More plausible would be the right of a child not to be caused by negligent or willful acts of its mother to be born so seriously defective that its life is not worth living, one plausible basis for wrongful life suits. Only a very few such cases have survived on appeal, but they could be supported in the courts or state legislatures by very plausible arguments. Although this limited right would provide a remedy for the most serious harms pregnant women inflict upon their offspring, it would need to be supplemented by other fetal rights to prevent other serious harms. In any event, there are several legal fetal rights that could plausibly be proposed as solutions to the social problem of the irresponsible actions of pregnant women.

Second, would these fetal rights be conditional on birth or not so conditioned? The legal rights of the unborn have traditionally been conditional on birth, and there are good reasons to continue this practice. In court cases where physicians have sought permission to impose blood transfusions or caesarian sections upon pregnant women, the predictions have often turned out to be highly unreliable. And the vast majority of babies born of mothers who have used illegal drugs during pregnancy have been born unharmed. To condition fetal rights upon live birth would limit legal intervention to those cases where it might be justified because serious harm has actually occurred. Moreover, it would respect parental discretion concerning the treatment of the unborn child just as our law now respects it after birth. The importance of giving parents, who usually love their children and are in the best position to judge the special circumstances of each case, wide but not unlimited discretion was recognized in *Prince v. Massachusetts*.[25] To confer new fetal rights holding against the mother that could be exercised before birth might give physicians and judges, most of whom are male, too much power to violate the pregnant woman's human and constitutional rights to liberty, privacy, and bodily integrity.

On the other hand, legal action postponed until after birth will often be too little and too late. It will be too late because the child will already have been harmed by the mother who has violated her maternal duties during pregnancy. If at all possible, fetal rights should be defined in a way that will permit legal intervention early enough to prevent the most serious harms irresponsible mothers inflict upon their offspring. After

---

[24] *Smith v. Brennen*, 157 A.2d 497 at 503.

[25] *Prince v. Massachusetts*, 321 U.S. 158 (1944) at 166.

all, these harms constitute the social problem and these proposed fetal rights are intended to solve. Moreover, fetal rights conditional on birth holding against the mother could at best provide only a grossly inadequate remedy for these harms. In medical malpractice cases, the victim can often sue for large amounts of compensation typically paid by means of medical malpractice insurance. But very few mothers have deep pockets and it would be counterproductive to require pregnant women to expend any significant portion of their limited resources for the purchase of parental malpractice insurance. Their money, like their time and energy, should be devoted to the care of their unborn children. Hence, there is little to be gained and much to be lost by postponing legal action in the name of the child until after birth.

One might, I suppose, advocate a combination of these two alternatives. Our legal system might recognize a very few fetal rights that are not conditional on birth. These should be defined very narrowly to cover only those cases where the medical diagnosis and prognosis are highly reliable, the predicted harm most serious, the conduct of the pregnant woman most irresponsible, and prenatal intervention reasonably effective. Fetal rights defined in this way would be justified both because they would enable third parties to intervene early enough to prevent very serious harms and because their scope would be restricted enough to protect the moral and constitutional rights of responsible mothers. In cases where the risk of harm to the unborn child is somewhat less and the reasonableness of the mother's actions is debatable, any new fetal rights should be conditional on birth.

Third, who could be empowered to take what legal action, before or after birth, in the name of the child? Because the fetus or neonate lacks the physical and psychological capacities necessary to perform any genuine action, neither could exercise its own legal rights. But the legal fiction of fetal or neonatal agency could easily enable others to act in its name. Several possibilities spring to mind. (1) The father or other close relative might be empowered to consent to fetal therapy in spite of the mother's refusal or to petition a court to order the pregnant woman to cease and desist from activities harmful to the fetus. After birth, he could be empowered to sue the mother for compensation for injury to the child or to petition the court for sole custody of the child to remove the child from the harmful influence of the mother. (2) The attending physician, already empowered to petition the court for permission to treat the fetus in spite of the pregnant woman's refusal, could exercise this power more frequently to prevent medical injury or to treat fetal diseases or disabilities *in utero*. (3) A child protection agency, acting on a complaint of a relative, physician, or social worker, could be empowered to petition a court to order the mother to cease and desist from activity harmful to the fetus or to transfer custody from the mother to some other party. Before birth this could be only a partial and temporary transfer of custody because the child cannot be physically removed from the pregnant woman alive before it is delivered, either naturally or by caesarian section. After birth, the child could be removed from the mother's home and placed in a foster home or public institution. (4) Pursuant to several of the above legal actions, the court might appoint a guardian *ad litem* for the unborn child. The authority of any such guardian might be very limited, perhaps to consent to or refuse medical treatment for the fetus, or broader to oversee the entire care of the fetus and ensure that this care is in the best interests of the unborn child. This authority would be conferred by the state acting in its capacity of *parens patriae*. None of these alternatives are a radical departure from existing law; each is simply an extension of legal powers available today.

Finally, what would be the most plausible proposal to extend the legal rights of the unborn child? (1) A narrowly defined right of the fetus to prenatal medical care should become, if it is not already, unconditional on birth. Either the physician or a close relative should be empowered to petition a court to require a pregnant woman to submit to such care when this care is urgently needed by the unborn child but only when this treatment would not impose undue harm or risk of harm upon the mother. In most cases, it would be better for the attending physician to take legal action, in consultation with the father or other close relatives, because the physician is in the best position to know how urgently the fetus needs treatment before birth and how great the medical risk would be for the pregnant woman. Also, family harmony would usually be seriously damaged where a pregnant woman forced to submit to unwanted treatment by the legal action of another family member.

(2) A narrowly defined right of the fetus not to be abused in any of a very few specified ways should be unconditional on birth. This could best be enforced by a child protection agency acting on a complaint by a social worker, guardian *ad litem*, or close relative of the pregnant woman. However, courts should issue cease and desist orders only in cases when the mother has already acted in some very irresponsible way and there is a reasonable prospect of ensuring compliance with the court order.

(3) It might well be useful to recognize more extensive legal fetal rights holding against the pregnant woman to all necessary care and not to be abused in any serious manner. The content of these rights should be defined fairly broadly but as specifically as possible. These more general fetal rights should be conditional on birth in order to ensure that they could support legal action only when the child has actually suffered harm and when a court will be able to assess the kind and degree of harm accurately and determine the kind of remedy that will be appropriate. Further details of these three suggestions, if they are found convincing, will need to be worked out by the courts or by state legislatures.

7.4. Objections and Replies

As one would expect, the various proposals for the introduction of new fetal rights into our legal system have occasioned strenuous objections. Several of these should be taken very seriously. First, these proposed fetal rights would interfere with parental discretion and disrupt family harmony. United States law has traditionally accorded parents great latitude in deciding how to care for their children. Such parental discretion was even recognized as a constitutional right in *Prince v. Massachusetts.*[26]

> It is cardinal with us that the custody, care and nurture of the child reside first in the parents, whose primary function and freedom include preparation for obligations the state can neither supply nor hinder.... And it is in recognition of this that these decisions [*Pierce v. Society of Sisters*, 268 U. S. 510, and *Meyer v. Nebraska*, 262 U. S. 390] have respected the private realm of family life which the state cannot enter.

[26] *Id.* at 166.

The reasons for wide parental discretion include the facts that the parents are usually in the best position to judge the best interests of their child given its special circumstances, that they normally want what is best for their child, that they must bear the financial and other burdens such as providing special care of any mistaken decision, and that state intervention tends to be misplaced and ineffective. These reasons surely apply to the unborn child at least as much as to the child after birth.

But this parental discretion is not and should not be unlimited. As *Prince* in the next paragraph held, "the family itself is not beyond regulation in the public interest."[27] Thus, child neglect and child abuse statutes are not unconstitutional and are morally justified, as long as they are not formulated in overly broad terms. Similarly, some of the proposed fetal rights would not unduly infringe the parental discretion of the mother provided they were defined narrowly enough so that their application would be limited to cases where the action of the pregnant woman is clearly unreasonable and threatens very serious harm to the unborn child.

Traditionally, our courts have accepted the parent–child tort immunity doctrine according to which the child lacks the legal power to sue her parents. As Justice Romiti, dissenting, observed in *Stallman v. Youngquist*, "Parental immunity is founded on the firm principle that the law should not intrude into parent–child interactions which occur in the context of and arise from the family relationship."[28] In addition to the reasons that support the recognition of parental discretion, the doctrine of parent–child immunity has been justified as essential to family harmony. To empower children to take legal action against their parents would transform what is normally a supportive relationship of parent to child into an adversarial one. This could hardly be in the best interests of the young child who is thoroughly dependent upon her parents who are primarily responsible for meeting all of her vital needs. Because the unborn child is even more dependent upon her mother as long as she is in utero, it is even more urgent to avoid alienating her only source of sustenance and parental care. Subsequently, she will need to be cared for within an intact and loving family to ensure her healthy emotional development.

However, Justice Johnson, who delivered the opinion of the court in *Stallman,* noted

> Other jurisdictions which have abolished the parent–child tort immunity doctrine have met the family disruption argument by reasoning that the injury itself and not the consequent suit is the factor which may upset the family unit. The widespread use of liability insurance mitigates against the possibility that such suits disrupt the domestic peace or deplete the family's financial resources.[29]

I do not find this reasoning completely convincing. Although it may be true that the injury to the child originally upset the family unity, any legal suit brought by the child against her mother would further disrupt the domestic peace and often to a much higher degree. This is another reason why any new fetal rights should be defined narrowly enough so that their application would be limited to cases where there is an urgent need to prevent serious harm to the unborn child and the least justification for the threatening actions of the pregnant woman. But to permit grossly irresponsible mothers to inflict very serious

---

[27] *Id.* at 166.
[28] *Stallman v. Youngquist*, 129 Ill.App.3d 859 at 865.
[29] *Id.* at 865.

harms upon their unborn children is too high a price to pay for what is likely to be only a temporary harmony between the mother and her offspring.

Second, the introduction of new fetal rights intended to protect the child against harmful actions of her mother would be counterproductive. For one thing, any possible enforcement of these rights would be ineffective. Consider the very wide range of interventions John A. Robertson, one of the most respected advocates of fetal rights, contemplates.

> If resolution of the fetal–maternal conflict in *Raleigh-Fitkin* and *Jefferson* in favor of the near-term fetus over the mother's interest in bodily integrity was correct, then far-reaching intrusions on the mother's body and freedom of action for the benefit of the unborn child may legitimately follow. Women, for example, might then be forced to deliver by cesarean section. They may also be prohibited from using alcohol or any substances harmful to the fetus during pregnancy, or to be kept from the workplace because of toxic effects on the fetus. They could be ordered to take drugs, such as insulin for diabetes, medications for fetal deficiencies, or intrauterine blood transfusions for Rh factor. Pregnant anorexic teenagers could be force-fed. Prenatal screening and diagnostic procedures, from amniocentesis to sonography or even fetoscopy, could be made mandatory. And, in utero surgery for the fetus to shunt cerebroventricular fluids from the brain to relieve hydrocephalus, or to relieve the urethral obstruction of bilateral hydronephrosis could also be ordered. Indeed, even extra-uterine fetal surgery, if it becomes an established procedure, could be ordered, if the risks to the mother were small and it were a last resort to save the life or prevent severe disability in a viable fetus.[30]

The surveillance and control of almost every aspect of the life of pregnant women needed to prevent this array of potential harms to the fetus would be as impossible as it would be repugnant.

More specifically, how might our legal system attempt to enforce fetal rights not conditional on birth? Giving custody of the unborn child to a guardian *ad litem* would confer very little real control over the child's welfare as long as the unborn child remains *in utero* and is entirely dependent upon sustenance from and vulnerable to injury from the body of the mother. A court order to cease and desist from using alcohol or illegal drugs would be futile when the mother is addicted, and an order to cease and desist from engaging in strenuous exercise or sexual intercourse could be enforced only by a degree of surveillance unavailable to the police. A court order to follow the advice of one's physician or to consume a diet adequate to nourish the fetus would be similarly impossible to enforce throughout a woman's pregnancy.

Fetal rights conditional on birth are by definition much less likely to prevent harm to the fetus because they could support legal action only after the injured child has been born. Nor would they be capable of providing adequate remedies for the harm suffered. A court order requiring remedial medical care will seldom be necessary, once the mother sees the plight of her child, and would often be unavailing because most mothers will lack the financial resources to pay for the very expensive care needed by their children. Suing the mother for compensation would seldom serve any purpose both because most

---

[30] Robertson, *The Right to Procreate & In Utero Therapy*, 333, 357–359.

mothers do not carry insurance against this sort of liability and because without insurance this would simply divert funds the mother would need to provide remedial care for her child. Taking custody away from the mother usually deprives the child of the care of the person best able to provide for its needs and sometimes transfers the vulnerable child to the custody of neglectful or even abusive social agencies or foster parents. This is not to say that introducing new fetal rights would never protect unborn children from harm, but a realistic study of past experience with similar legal measures suggests that this would be much less often than the proponents of fetal rights imagine.

At the same time, the enforcement of the proposed fetal rights would often harm those they were intended to protect. Pregnant women addicted to alcohol or illegal drugs would often avoid prenatal medical care for fear of losing custody of their babies and control of their pregnancy. And it is not only addicts whose unborn children will be harmed by the lack of necessary prenatal medical care. Women whose religious convictions rule out threatened medical treatments such as blood transfusions or those who object to abortion on moral grounds and fear that they might be forced to abort a seriously defective fetus might well avoid entanglement with the medical establishment. Many women who are already suspicious of the prejudice and arrogance of physicians, witness the proliferation of medical malpractice suits, will be reluctant to subject themselves to treatment by potential legal adversaries. Pregnant women who do seek and obtain continuing prenatal treatment may no longer trust their physicians sufficiently to reveal to them medical problems that might elicit coercive intervention by the law. Attempts by the father or other close relative to enforce fetal rights would disrupt the family harmony necessary for the supportive environment that would enable the child to grow up and develop fully and normally. And to transfer custody to a foster home or social institution will frequently harm the child by depriving her of intimate personal relationships and sometimes exposing her to serious neglect or even abuse. On balance, then, introducing new fetal rights into our legal system would be counterproductive because it would do very little a protect children from harm and would sometimes inflict new harms upon them.

To my mind, this is a serious objection that suggests caution rather than the rejection of every proposal to expand fetal rights. In particular, it will be important to reflect carefully and in detail upon how any new fetal right would work in practice. Enacting a fetal right by statute or recognizing one in an opinion of some court will in and of itself do nothing to prevent harm to children. Legal realism implies that what really matters is how the expanded law of fetal rights actually functions. A new fetal right should be introduced only if there is reason to believe that it can be effectively enforced and that the normal procedures for enforcing it will not result in unduly harmful side-effects. To this end, these rights must not be defined in overly grand and general terms inviting their abuse, and the power to take legal action to protect fetal rights must be conferred only upon those who are likely to do so most effectively and with restraint.

Third, the proposal to protect children by the introduction of new fetal rights is misguided. Its advocates misconceive the very real social problem arising from the conduct of pregnant women who neglect or abuse their unborn children. They imagine that the primary cause of such conduct is the moral irresponsibility of many individual women. But the primary causes of their conduct lie in our inadequate social institutions. Unemployment, lack of affordable childcare, and poverty prevent many mothers from consuming a diet that would provide adequate nutrition to their fetuses. Women often become addicted to alcohol or illegal drugs because of the desperate circumstances in which they feel themselves trapped and cannot escape their addiction because of the lack of substance abuse programs. Worse yet, most of the few programs that do exist do not permit pregnant women to enroll. Teenagers become anorexic and might need to be force fed only because of bodily ideals imposed upon them by commercial advertising and peer pressure. And pregnant women most often deny their unborn children the prenatal care they need, frequently unaware of that need, because they cannot afford the high cost of modern medicine but still do not qualify for Medicaid. One of the most knowledgeable advocates of fetal rights, John A. Robertson, admitted:

> The most effective measures to prevent prenatal harm to offspring are likely to be noncoercive policies that educate women and men about prenatal risks and provide services and treatment essential to offspring welfare. In many cases, children are injured by prenatal conduct because of the parent's unawareness of the danger, or lack of access to the treatment or other services needed to prevent the harm.[31]

Arguably, proposals to protect children by introducing new fetal rights are misdirected because they fail to realize that the primary cause of harm to human fetuses is not individual conduct as much as the social context in which pregnant women live and act.

I know of no convincing measure by which one could distinguish between the primary and the secondary causes of harm to unborn children. These serious harms arise from the conduct of individual women acting or failing to act as would be in the best interests of their unborn children in considerable measure because of the social context in which they live. But just as the causation of fetal harm is not either individual conduct or social context independently, so no adequate solution can be simply either introducing new fetal rights or reforming our social institutions. Both are required to prevent the birth of children with avoidable diseases or defects. Hence, new fetal rights may well be justified as one, but only one, part of the solution to the problem of harm to unborn children.

Some might charge me with naïve idealism. Granted that any complete solution to the complex problem of harm to the unborn would require the introduction of new fetal rights, it is unrealistic to imagine that any such perfect solution is practicable. Janet Gallagher argues

> As Katha Pollitt points out ["Fetal Rights: A New Assault on Feminism," *The Nation*, March 26, 1990, p. 1], "The focus on maternal behavior allows the government to appear to be concerned about babies without having to spend any money, change any priorities, or challenge any vested interests." But this insistence on pointing the finger of blame at individual women is an exercise in collective bad faith, a social self-deception which rationalizes our passivity toward the genuinely horrifying living conditions confronting many poor women.[32]

Thus given the realities of practical politics, the introduction of fetal rights would do little or nothing to protect unborn children because that would not only fail to change

[31] Robertson, *Children of Choice* 181.
[32] Janet Gallagher, "Collective Bad Faith: "Protecting" the Fetus". *Reproduction, Ethics and the Law*, Joan C. Callahan (ed.) (1995), pp. 343, 352.

the social context within which pregnant women must act but also provide an excuse for leaving our harmful social institutions intact.

Far be it from me to underestimate the public inertia that hinders the improvement of social institutions. But this inertia exists whether or not new fetal rights are introduced. The best strategy is probably to attack on all fronts at once, both to introduce a few new fetal rights and at the same time to make every effort to reform our social institutions when they encourage or even force pregnant women to lead their lives in ways that seriously injure their offspring. However, it would be unwise to extend fetal rights substantially without significant institutional reform, for under those circumstances the new rights would do more to harm pregnant women than to protect their offspring.

Fourth, the introduction and enforcement of new fetal rights would infringe several fundamental moral and legal rights of pregnant women. (1) They would often deny their right to bodily integrity and the right to refuse medical treatment this implies. Joelyn Knopf Levy reminded us that although physicians are quick to accept medical uncertainty as justification for their errors, they are less quick to recognize its implications for patient self-determination. She went on to note that it almost appears that the more a patient resists a physician's advice, the more dire the physician's predictions become.[33] Hence, they will often be disposed to appeal to the courts to enforce any fetal right to medical treatment.[34] And *In Re A.C.* indicates that the courts will often permit the physician to treat a pregnant woman in spite of her refusal. "The state's interest in protecting innocent third parties from an adult's decision to refuse medical treatment, however, may override the interest in bodily integrity."[35]

It is true that any fetal right to necessary medical treatment or to all necessary care holding against the mother could, where the pregnant woman to refuse such care, override the woman's right to refuse medical case in some cases. It is probably also true that physicians take their duty to provide good medical care to the unborn child, and at least to do no harm to it, so seriously that they tend to somewhat overestimate the risk of harm to the fetus where it to be denied the recommended medical treatment. Still, the physician is presumably much better qualified than the pregnant woman to know both the burden of this treatment on her and the potential harm to the fetus lacking it. Nor am I so sure that *In Re A.C.* really indicates that the courts will very often intervene to force unwanted treatment upon pregnant women. Because in that case the baby delivered by caesarian section died 2 hours later, that case serves more as a warning against than an invitation to court interventions. That the state's interest in protecting innocent third parties may sometimes override the right to bodily integrity does not imply that it will do so often or that when it does so, legal intervention would always be unjustified.

(2) Any new fetal rights would necessarily infringe the pregnant woman's right to the equal protection of the law. Dawn E. Johnsen made this point very clearly.

> Only women can suffer the great intrusions of such laws, for only women have the ability to bear children. Fetal rights laws would not only infringe on constitutionally protected liberty and privacy rights of individual women, they

---

[33] Joelyn Knopf Levy, "Jehovah's Witnesses, Pregnancy, and Blood Transfusions: A Paradigm for the Autonomy Rights of All Pregnant Women", *Journal of Law, Medicine and Ethics* 27 (1999), 171, 184.

[34] See also Cheryl M. Plambeck, "Divided Loyalties: Legal and Biomedical Considerations of Physician-Pregnant Patient Confidentiality and Prenatal Drug Abuse", *Journal of Legal Medicine* 23 (2002), 4–5, 27.

[35] *In Re A.C.*, 533 A.2d 611 (D.C. App. 1987) at 616.

> would also serve to disadvantage women as women by further stigmatizing and penalizing them on the basis of the very characteristic that historically has been used to perpetuate a system of sex inequality.[36]

Even if the courts attempted to balance the interests of the fetus against those of the mother equally, fetal rights would, by their very nature, deny pregnant women the equal protection of the law.

I do not believe that fetal rights, by their very nature, necessarily infringe the pregnant woman's right to the equal protection of the law. Granted that fetal rights holding against the mother will sometimes ground legal intrusions into her life. But she is not the only party whose actions may harm the unborn child. Fetal rights can and should be defined so that they also hold against the father, the physician, and any others in a position to injure the fetus. To be sure, they may more often occasion intrusions into the life of the pregnant woman than into other lives because her unique physiological relation to the fetus puts her in a special position to harm, or to benefit, the unborn child. But this does not deny her equal protection of the law any more than the fact that only physicians or other medical practitioners can be sued for medical malpractice violates their right to equal protection.

(3) Any fetal right not to be injured in a way that would cause the child to be born seriously defective, and some such right would be needed to prevent the very worst harms to unborn children, and would infringe the pregnant woman's right to privacy regarding abortion.[37] Many medical malpractice suits hinge on the failure of physicians to give the mother an opportunity to abort. This suggests that a child could sue the mother for failure to abort when the mother had been informed that her offspring would be born with seriously impaired life prospects.

> To avoid the conclusion that wrongful life actions restrict a woman's procreative freedom, one must show that she owes no legal duty to abort her impaired fetus. But any suggestion that she owes no duty to the fetus forces one to face the dilemma of whether she wrongs the fetus by bringing it to term. If there is no wrong in her bringing it to term, then one wonders why the physician is said to wrong the fetus by failing to inform the mother of its defects.[38]

It is not only wrongful life suits that would force some women to have an abortion against their wills. Dorothy Roberts observed that "Women who are punished for drug use during pregnancy, then, are penalized for choosing to have the baby rather than having an abortion. It is *the choice of carrying a pregnancy to term* that is being penalized."[39] But it is a woman's choice of whether *or not* to have an abortion that is protected by her constitutional right to privacy. Hence, any fetal right forcing a woman to have an abortion violates her constitutional right to privacy just as much as a statute prohibiting her from doing so.

---

[36] Dawn E. Johnsen, "The Creation of Fetal Rights: Conflicts with Women's Constitutional Rights to Liberty, Privacy, and Equal Protection", *Yale Law Journal* 95 (1985), 599, 620.
[37] See, Plambeck, *Op. Cit.*, pp. 17–18.
[38] George Schedler, "Women's Reproductive Rights", *The Journal of Legal Medicine* 7 (1986), 357, 366.
[39] Dorothy Roberts, *Killing the Black Body* (1997), 181 italics in original.

I believe, although other conscientious individuals will not, that it is sometimes morally wrong for a woman to have an abortion or, when a child would be born into an intolerable life, to refuse to abort her fetus. Nevertheless, I believe that the pregnant woman has the moral right and should have the legal right to make the decision of whether to abort or carry her fetus to term because this is a private decision. Therefore, I believe that any new fetal rights should be limited so that they would not imply any legal obligation of the mother to abort against her will. This does not, however, rule out the possibility or even the desirability of introducing fetal rights with this restriction.

(4) The various proposed fetal rights would infringe the pregnant woman's human and constitutional right to liberty by imposing far-reaching constraints upon her.

> Treating pregnancy as a conflict between maternal and fetal rights leads inevitably down a slippery slope. Prosecutions of pregnant women cannot rationally be limited to illegal conduct because many legal behaviors cause damage to developing babies.[40]

Dawn E. Johnsen described the potential expansion of legal coercion required to enforce fetal rights and thus the many ways they would deny her right to liberty of action.

> Given the fetus's complete physical dependence on and interrelatedness with the body of the woman, virtually every act of the pregnant women has some effect on the fetus. A woman could be held civilly or criminally liable for fetal injuries caused by accidents resulting from maternal negligence, such as automobile or household accidents. She could also be held liable for any behavior during her pregnancy having potentially adverse effects on her fetus, including failing to eat properly, using prescription, nonprescription and illegal drugs, smoking, drinking alcohol, exposing herself to infectious disease or to workplace hazards, engaging in immoderate exercise or sexual intercourse, residing at high altitudes for prolonged periods, or using general anesthetic or drugs to induce rapid labor during delivery. If the current trend in fetal rights continues, pregnant women would live in constant fear that any accident or "error" in judgment could be deemed "unacceptable" and become the basis for a criminal prosecution by the state or a civil suit by a disenchanted husband or relative.[41]

If this prediction is even roughly accurate, then it constitutes a very strong objection to any proposal to introduce new fetal rights into our legal system.

In theory, there is a short and easy reply to this objection. Not even fundamental moral or legal rights are absolute or unlimited. In particular, the individual's right to liberty does not permit one to cause avoidable unnecessary harm to others. In practice, however, it is very difficult to draw the line between justified liberty and wrongful license. This will be especially true given the wide range and variety of maternal actions that are potentially harmful to the fetus. But however difficult this may be, state legislatures and our courts have been reasonably successful in defining the boundaries of our fundamental legal liberties in morally justified ways. I am hopeful that if fetal rights that could support

---

[40] Lynn M. Paltrow, "When Becoming Pregnant Is a Crime", *Criminal Justice Ethics* 9 (1990), 41, 42.
[41] Johnsen, *Supra* note 36, 605–607.

prenatal legal actions are defined very narrowly and fetal rights conditional on birth are introduced only gradually and with caution, they need not unduly limit the liberty of pregnant women.

Finally, (5) new fetal rights would deny many women procedural due process. Janet Gallagher has surveyed the recent judicial decisions in support of the forced treatment of pregnant women and concluded that their authority is undermined by the lack of due process.

> If the pregnant woman is represented by counsel at all, it is virtually always by a last minute court appointee operating under an overwhelming caseload and demanding time constraints. The parties ruled against are seldom motivated or well situated to fully brief or argue the complicated woman's rights or fetal status issues. Mootness, or the losing party's disinterest in (and practical inability to pursue) full scale litigation insulate decisions from appellate review. Amicus briefs from groups with special expertise and interest in the broader issues raised by a specific case can seldom be solicited under the timetables imposed by nature.[42]

Arguably, any expansion of fetal rights would increase the frequency with which the right to procedural due process of pregnant women would be infringed.

No doubt several of the early cases in which the courts have permitted the medical treatment of pregnant women against their wills were tried under conditions that did not provide those women adequate due process of law. But this is because most of these were responses to what were perceived to be medical emergencies late in pregnancy that required a quick decision to save the life of the unborn child. But the proposed fetal rights would apply to a much broader range of cases, many of which would concern medical treatment earlier in a pregnancy and allow more time for a full and fair hearing before the court. This does not mean that the expansion of fetal rights poses no real threat to the mother's right to due process of law, but it does suggest that any new fetal rights could be defined in a way that would minimize this threat. In sum, I cannot, nor would I wish to, deny that these potential infringements of five fundamental rights of pregnant women constitute a serious objection to recent proposals to introduce new fetal rights into our legal system. But I believe that they require caution in acting on these proposals, not their complete rejection.

Fifth, even if new fetal rights were justified in principle, their introduction into our legal system would be unjustified because in practice they would be applied in an unjust manner. Our social institutions would discriminate against African-Americans, the poor and women when enforcing fetal rights holding against pregnant women. Dorothy Roberts harshly criticizes the application of child abuse statutes to pregnant women in South Carolina.

> The reason Black women are the primary targets of prosecutors is not because they are more guilty of fetal abuse. A Study of twenty-four hospitals conducted by the South Carolina State Council on Maternal, Infant, and Child Health in 1991 found that high percentages of pregnant women were abusing

[42] Janet Gallagher, "Prenatal Invasions & Interventions", *Harvard Women's Law Journal* 10 (1987), 9, 48–49.

> marijuana, barbiturates, and opiates—drugs used primarily by white women. MUSC's own record showed that drug use among pregnant women was evenly distributed among white and Black women. Yet nearly all of the women the hospital reported to the solicitor were Black.... This discriminatory enforcement is a result of a combination of racism and poverty. Poor women, who are disproportionately Black, are in closer contact with governmental agencies, and their drug use is therefore more likely to be detected. Black women are also more likely to be reported to governmental authorities, in part because of the racist attitudes of health care professions.[43]

What is true of South Carolina is probably in large measure true of most other states.

There is similar evidence of discrimination against women in the enforcement of existing fetal rights.

> There has also been criticism of the glaring gender discrimination inherent in focusing "fetal protection" measures on pregnant women while ignoring male behavior and substance abuse. Pamela Rae Stewart, for example, was charged with having disregarded doctor's orders by having sexual intercourse with her husband. No charges were ever brought against the husband, even though he had also been aware of the doctor's instructions. In one Wyoming case, a pregnant women who had fled to a hospital emergency room after being beaten by her husband was arrested and jailed on child endangerment charges when tests revealed that she had been drinking. Critics of the widespread use of drug screening tests on newborns to seize custody of children on the grounds of maternal neglect or abuse point out that reliance on such tests fails to identify drug use by fathers. The insistent emphasis on individual maternal responsibility for fetal and child well-being is not only sexist, but ineffective public policy. It ignores the reality that men have an impact on the well-being of their children—before, during, and after pregnancy.[44]

Given the discriminatory enforcement of existing fetal rights, it would presumably increase the injustice of our legal system to introduce any new fetal rights.

In particular, it might well invite discrimination against women to confer any legal right to necessary medical treatment on the fetus. In a 1990 report published in the *Journal of the American Medical Association*, the AMA asserts that a pregnant woman has a moral duty regarding her fetus, but observes that this does not necessarily impose any similar legal duty. Although it lists several reasons why resorting to the courts is usually undesirable, it also suggests that there may be exceptional circumstances in which a physician is ethically permitted to seek a court order to treat the mother. Joelyn Knopf Levy argued that the medical profession is excessively prone to impose treatment upon pregnant women.

> The AMA and ACOG have not elaborated on those circumstances where court intervention is justified, but a 1987 survey of teaching hospitals in forty-five states provides an indication of those situations that impel physicians to seek court orders. The survey, conducted by Veronika Kolder, Janet Gallagher, and

---

[43] Roberts, *Supra* note 34, 172.

[44] Gallagher, *Supra* note 29, 358.

> Michael Parsons, reveals a startling proclivity among obstetricians for forcing treatment on unwilling pregnant women. The survey, the only one of its kind reported in the medical literature, was sent to heads of maternal–fetal medicine fellowship programs and directors of maternal–fetal medicine divisions in obstetrics/gynecology residency programs. Forty-six percent of these physicians believed that pregnant women who refused medical advice, thereby endangering the life of the fetus, should be detained in hospitals or other facilities to ensure (that is, force) compliance. Forty-seven percent agreed with court decisions ordering cesarean sections and supported the extension of such precedent to other procedures such as intrauterine transfusion. Twenty-six percent advocated state surveillance of women in the third trimester who "stay outside the hospital system."[45]

The fact that the vast majority of these physicians were males strongly suggests that they would enforce a fetal right to medical treatment holding against the mother in a highly discriminatory manner.

There is, I fear, a great deal of truth in this objection to the recent proposals to introduce new or expanded fetal rights into our legal system. Our social institutions do far too often discriminate against African-Americans, the poor, and women. Hence, old fetal rights have been and new fetal rights probably would be applied against many pregnant women unjustly. This will suggest to some that we ought to postpone the extension of fetal rights until there is less discrimination in our society. I believe, however, that it would be better to move forward cautiously in the gradual introduction of new fetal rights at the same time that we attempt to reform our social institutions to render them more just. Indeed, as more and more African-Americans and women enter the professions of law and medicine, a process that is already in progress, our courts and healthcare institutions will presumably become less discriminatory. In the long term, the more difficult problem will probably be to reduce significantly discrimination against the poor, especially African-Americans and women living in poverty.

Although discriminatory enforcement of the law is always morally objectionable, it is not necessarily a sufficient reason to condemn laws that protect, even imperfectly, the vulnerable members of our society. A few comparisons may be illuminating. It seems clear that in our metropolitan areas, police protection is not provided equally to all segments of the population. Typically, there is more and better protection against theft, bodily injury, and death in the wealthier suburbs than in the urban cores where African-Americans, members of other minority groups and poor whites predominate. This is surely objectionable, but it would not in and of itself justify refusing to introduce new laws to provide new forms of protection for our citizens. Again, statutes against the physical or sexual abuse of young children are often applied in a discriminatory manner, but this would not be a sufficient reason to oppose the extension of these laws to cover the psychological abuse of children. Finally, a woman's tort right not to be battered may well be applied more often against African-American and poor men than against more affluent male batterers, but it would still be a moral improvement to extend any law against wife-battering to cover unmarried partners of violent males.

---

[45] Levy, *Supra* note 30, 177.

7.5. Conclusion

Thus, there are very serious objections to the introduction of new or expanded fetal rights into our legal system. Although I believe that I have given a plausible rebuttal for each of these five objections, I remain gravely concerned about their cumulative force. Let us, therefore, reconsider the proposals for extending old fetal rights and introducing new fetal rights. The usual argument for fetal rights holding against the mother seems simple at first glance. (1) Pregnant women have a moral duty not to harm their unborn children and to submit to fetal therapies needed to preserve the health of children to whom they will give birth. (2) Pregnant women will sometimes violate these moral duties with very serious consequences for their offspring. Therefore, the law should extend old fetal rights and recognize new fetal rights in order to enforce these maternal duties and thereby prevent avoidable harms suffered by the children who would otherwise be born with serious medical problems and be unable to develop into fully functioning citizens of our society.

The first premise in this reasoning is roughly correct. Section 7.1 above shows that pregnant women do have moral duties regarding their fetuses, but that these duties are not quite as broad as is often assumed. Their moral duty not to abuse their unborn children does not require them to avoid every action that might cause harm, but only those willful or negligent acts that would seriously harm the child who will be born. And their duty of care requires only that they provide for the basic needs of their fetuses. Hence, it implies only a moral duty to submit to medical care necessary for basic good health and normal abilities for the unborn child, not every treatment that might contribute to perfect health and enhanced capacities. Also, it does not exclude postponing treatment until after birth when to do so will not seriously harm the child.

There are other qualifications that one should bear in mind. Moral obligations presuppose that the duty-bearer can perform the required act and can do so without excessive sacrifice. Accordingly, women genuinely addicted to drugs or alcohol or tobacco have no duty to stop using these harmful substances immediately upon becoming pregnant, although presumably they do have a duty to seek assistance in overcoming their addiction. Similarly, neither mothers nor fathers have any moral duty to provide medical care they genuinely cannot afford. And because moral duties are also undermined by excessive sacrifice, pregnant women may have no moral obligation to submit to prenatal medical treatment when this seriously burdens any of their fundamental moral rights, including their right to bodily integrity. Finally, it is important to remember that pregnant women are not the only persons who have moral duties not to harm and even to care for unborn children. The recent focus on maternal–fetal conflict should not obscure the fact that others, especially fathers and physicians, have similar duties regarding human fetuses. Therefore, if the maternal duties of pregnant women justify legal fetal rights holding against mothers, the duties of others justify defining these fetal rights so that they hold against others also.

The second premise in this argument is also true. Section 7.2 above describes several reasons to believe that pregnant women will sometimes, perhaps often, violate these maternal duties with very serious consequences for their offspring. But how often will they violate their maternal duties to their unborn children? Will it be often enough to constitute a very serious social problem or only occasionally and under extenuating circumstances? And when would pregnant women choose to act in ways that might cause serious harm to their unborn children? Will it be in the kinds of cases where some fetal right could prevent or at least remedy the potential harm or in situations to which fetal rights would be inapplicable or ineffective? Until we know the answers to these questions, we cannot know whether this second premise provides a sound reason for extending old fetal rights and introducing new ones.

Unfortunately, these questions have been largely ignored in the heated debates concerning maternal duties and fetal rights. Those who have advocated expanded fetal rights have described an impressive variety of maternal actions or failures to act that might, in the light of modern medical knowledge, harm their unborn children and have usually proposed an equally wide range of fetal rights that might prevent these harms. Opponents have described a wide variety of ways in which many of these proposed fetal rights would be either useless or seriously objectionable. To my mind, it is futile to continue the debate on this level of generality. It is a mistake to attempt to balance the pro and con arguments wholesale because their weights vary from case to case. What is needed is a piecemeal consideration of the very different situations to which some specific fetal right might be applicable and to consider each kind of case on its own merits.

Are there specific kinds of cases where some pregnant women will violate their maternal duties in a way that probably will cause serious harm to their offspring and some fetal right could apply without being subject to the most serious objections of those who oppose fetal rights? Well, there is evidence to show that considerable numbers of pregnant women do violate their moral duty not to abuse their unborn children by willfully or negligently using harmful substances such as illegal drugs, alcohol, and tobacco. Previous attempts to prevent such substance abuse have been both ineffective and objectionable. Child abuse statutes have in some states been extended to apply to fetuses, either to impose criminal penalties upon women who have used illegal drugs during their pregnancy or to remove children from their custody soon after birth. But punishing the mother has done little or nothing to reduce the exposure of unborn children to illegal drugs, and removing the child from the custody of her mother often deprives that child of the person best able to care for her needs. Worse yet, both measures have been applied in a very discriminatory manner. It might be possible, however, to confer a fetal right under tort law not to be willfully or negligently exposed to excessive amounts of harmful substances such as illegal drugs, alcohol, or tobacco. Any such right should hold, not only against the mother, but also against the father and other members of the family. It could enable physicians or social workers to petition a court for an order that the pregnant women cease and desist from the use of such harmful substances and submit to periodic monitoring to ensure her compliance. The court could also impose an order upon the father or another family member not to provide such substances to the mother or encourage her to use them. Only experience could determine whether any such fetal right would be effective in protecting unborn children, but it is a more promising approach than those that have already been tried without success.

The maternal moral duty to care for one's unborn child also implies a duty to provide medical care for one's fetus when this is necessary for the basic health and normal abilities of the child who will be born. But some pregnant women will refuse their consent to such treatment because they believe it to be inconsistent with their religious faith. Others will refuse consent because they are insufficiently adverse to risks and are willing to gamble on the health of their child, or because they have unreasonable fears of medical technology, or simply because they prefer the natural to the technological. Presumably, it is better in general to give parents wide discretion regarding the care of their children, born or unborn;

but there are limits beyond which they should not be permitted to expose their children to harm. The courts have already recognized a right of unborn as well as born children to necessary medical care. Perhaps, this common law right ought to be further developed on a case-by-case basis. Past cases in which the courts have ordered caesarian sections sound a cautionary note. Judges have often overestimated the need for this major surgery and underestimated the burdens thus imposed on women who have refused to give their free informed consent to it. But *in utero* surgery performed by fetoscopy is probably less intrusive than a caesarian section, and a court might be justified in permitting it to correct bilateral hydronephrosis or obstructive hydrocephalus in spite of the objections of the mother. Even less objectionable would be to order an interuterine blood transfusion when the Rh factor of the mother is incompatible with that of her unborn child. Finally, the fetus might have a legitimate right that prenatal testing be conducted when there are indications of a risk of some serious defect. Although a woman might refuse her consent in order to avoid the need to choose whether to have an abortion, in some cases, prenatal therapy might make an abortion unnecessary. In any event, there are several plausible examples of cases where some new or extended legal fetal right might be justified. Hence, this second premise might well support the conclusion that such rights ought to be introduced into our legal system. Whether this conclusion really follows remains to be seen.

It does not follow immediately because the conclusion in the usual argument rests on two additional premises. One assumption is that (3) the proposed fetal rights would in fact prevent the threatened medical harms that would otherwise be suffered by the children to be born. Opponents object that fetal rights would in practice be ineffective in preventing the predicted medical harms. Although this is probably true of some of the fetal rights proposed by the more enthusiastic advocates, it is premature to judge that this would always be the case. In the plausible cases described above, the specified fetal rights might well prove effective in avoiding or at least reducing very serious medical problems for children who will be born.

The other additional assumption is that (4) the proposed legal fetal rights would serve to enforce the prior moral duties of pregnant women. Although some of the very broad fetal rights that have been proposed might call for actions over and beyond the demands of moral duty, those specified in the plausible cases described above probably fall well within the bounds of the moral obligations of pregnant women. And there may well be other fetal rights that would also serve to enforce some maternal moral duty of pregnant women regarding their fetuses.

Given all four premises, properly qualified, should we conclude that our legal system ought to extend some old fetal rights and recognize some new fetal rights? Well, if premise (3) that some of the proposed fetal rights would in fact prevent the threatened medical harms that would otherwise be suffered by the children to be born is true, then that conclusion is strongly supported. Normally, the fact that some provision of the law would prevent serious harm is a primary reason for introducing and enforcing that legal provision. Notice that at least two kinds of harms are at stake here. First, there are the medical diseases or defects suffered by the individual children to whom the pregnant women will give birth. Second, there are the harms of diverting scarce medical resources from more cost-effective purposes to providing the extraordinary care needed by these children and their inability to grow into fully functioning productive citizens. Hence, carefully selected and defined legal fetal rights would be justified by both the state's duty to protect its individual members from harm and by the state interest in public welfare.

Unless, of course, these fetal rights would have side-effects even more harmful than the harms they were intended to prevent. Opponents of fetal rights holding against the mother often allege that this would be true. For example, any right to prenatal genetic testing might motivate many pregnant women to hide from their physicians serious medical problems in their family history. And any right that would require a patient to undergo periodic testing for the use of illegal drugs or alcohol might discourage many patients from seeking or accepting adequate prenatal medical care. At this point, these and other alleged harmful side-effects of fetal rights are highly speculative. They suggest only that fetal rights should be expanded very gradually with a constant monitoring of their consequences, not that these rights could not be justified by their effectiveness in preventing serious harms. Presumably, then, the argument for the gradual selective extension of old fetal rights and recognition of new fetal rights is sound.

But if this straightforward argument from the prevention of harm to the justification of legal fetal rights is sound, then premise (4) would seem to be redundant. Why, then, do those who propose fetal rights typically insist that pregnant women have moral duties regarding their fetuses? What does the assumption that legal fetal rights would enforce prior moral duties contribute to the justification of these rights? Because I am unable to read the minds of the advocates of fetal rights, I will explore the hypotheses that come to my own mind.

One hypothesis is that those who propose new or extended legal fetal rights need this assumption to counter the objection that fetal rights would lead to the violation of several fundamental rights of pregnant women. Many jurists and moral philosophers believe that the assumption that these fetal rights will prevent more harm than they will cause cannot justify their introduction into our legal system because this is an appeal to utility and utility cannot override individual moral rights, especially fundamental rights. But if pregnant women have moral duties to their fetuses, then these might be thought to be logically correlative to moral rights of their fetuses. And one could then argue that the moral rights of the fetus override the moral rights of their mothers in a way that mere utility cannot.

If this is the proper way to interpret their argument, I find it unconvincing. Fetuses are not the kinds of beings that are capable of possessing moral rights. For reasons I have explained elsewhere, only a being that possesses moral agency could possess any moral right.[46] In any event, an appeal to the moral rights of the fetus is unnecessary. The real justification for legal fetal rights is grounded, not on preventing harms to or respecting the rights of fetuses, but primarily on preventing harms to and perhaps respecting the rights of those children who will subsequently be born.

Another hypothesis is that the moral duties of pregnant women would justify fetal rights holding against these women because legal fetal rights of this kind would enforce the prior moral duties of pregnant women and the law ought to enforce moral duties. But this hypothesis also fails to strengthen the argument for fetal rights. Although some, probably many, moral duties ought to be legally enforced, there are others that the law should ignore. Presumably, every promisor has a moral obligation to keep her promise. But the law ought not to enforce a trivial promise like one's promise to meet a friend

[46] Carl Wellman, *Real Rights* (1995), 137–145.

for lunch. And it would be morally objectionable for the law to attempt to enforce the promise of a parent to take his son to a baseball game because this would be intruding into the area of proper parental discretion in the care of one's child. Which moral duties, then, ought the law to enforce? Probably, either those duties the violation of which would cause serious harm to the some individual victim or those the violation of which would be seriously harmful to the public welfare. But if so, then this appeal to moral duties reduces to the appeal to the prevention of harm and adds nothing to the argument for fetal rights.

A third hypothesis is that advocates of legal fetal rights need to assume that pregnant women have moral duties regarding their fetuses in order to counter the charge that fetal rights would unjustly penalize pregnant women. Although any legal system ought to prevent harm to the members of its society, it is impermissible to do this by enforcing an unjust law. Even a very good end does not justify an unjust means. Opponents of fetal rights object that legal fetal rights would be unjust because they would place the entire burden of preventing medical harms to the unborn on mothers when fathers, other individuals, and society as a whole share this responsibility with them. Advocates of fetal rights may appeal to the moral duties of pregnant women to rebut this charge of injustice.

The reasoning may be that it is not unjust to burden pregnant women with the legal duties these fetal rights would impose because their enforcement against pregnant women would merely require them to do what they are already morally required to do. On this interpretation, the appeal to moral duties does contribute something to the argument. Moral duties are grounded on moral duty-imposing reasons, reasons both for a moral agent to act or refrain from acting in some specific manner *and* for others in society with that agent to react negatively should she act contrary to this reason.[47] Thus, the moral duties of pregnant women not to abuse their fetuses and to provide necessary medical care for them are grounded on reasons that would in some measure justify society in reacting negatively by imposing legal sanctions upon mothers who violate these duties. To this extent, the appeal to moral duties constitutes an additional independent justification for fetal rights.

But it does not entirely rebut the objection that legal fetal rights holding against the mother would be unfair to pregnant women. It is also necessary that these fetal rights be defined and applied justly. Thus, they should be defined so that they hold not only against mothers, but also against fathers, physicians, and other individuals in a position to harm or contribute to harming unborn children. And when applied, the courts must not demand greater sacrifices by pregnant women than the law requires of parents, mothers or fathers, of children who have been born.[48] And they must be applied in a way that does not discriminate against women as women, and especially against pregnant women who choose not to abort an unborn child with serious medical problems. Clearly considerable reform of our legal system will be required to eliminate every potential injustice that could arise from extended or new fetal rights. But this reform is already needed with or without the proposed expansion of fetal rights. And any residual unfairness of the penalties they would in practice impose on pregnant women must be weighed against the injustice of denying their offspring the equal protection of the law.

The argument as I have interpreted it is sound. The prevention of harm together with the moral duties of pregnant women regarding their fetuses would justify some expansion

---

[47] *Id.* at 49–59.
[48] Compare Thomas H. Murray, *The Worth of a Child* (1996), 100–103.

of old fetal rights and the introduction of some new fetal rights in our legal system. But it would do so only with the qualifications noted above, limitations required in order to meet the objections of those who oppose fetal rights. Consequently, this reasoning implies a less sweeping conclusion than that often urged by those who have recently advocated fetal rights holding against the mother.

More specifically, what does this reasoning imply about the most plausible fetal rights proposed in Section 7.3 above? (1) A narrowly defined right of the fetus to prenatal medical care holding against both parents should become unconditional on birth. Either the physician or a close relative should be empowered to petition a court to require a pregnant woman to submit to medical care when this care is urgently needed by the unborn child but only when this treatment would not impose undue risk of harm or excessive sacrifice upon the mother. It would be best for this right to be introduced by the courts. This would avoid the dangers of overly broad legislation because judicial decisions are tied to the specific facts of particular cases and allow the courts not only to expand the legal rights of the unborn, but also to limit fetal rights when the need arises.

(2) A narrowly defined right of the fetus not to be willfully or negligently abused by either parent in any of a very few specified ways should also be unconditional on birth. This might require state legislatures to modify family law statutes prohibiting child neglect or abuse to apply to unborn children. Such legislation could best be enforced by a child-protection agency acting on a complaint by a social worker, guardian *ad litem*, or close relative of the pregnant woman. However, courts should issue cease and desist orders only in cases where the mother or father has already acted in some very irresponsible way and there is a reasonable prospect of ensuring compliance with the court order.

However, I do not believe that (3) it would be useful to recognize more extensive fetal rights conditional on birth holding against the pregnant woman. Any such general rights would make possible and perhaps even invite violations of several basic moral and legal rights of pregnant women. Moreover, they would probably do little or nothing to remedy or relieve the suffering of children who have already been born with serious medical problems. Such rights would authorize parental malpractice suits, but these would not in practice have even the limited usefulness of medical malpractice suits because parents lack the deep pockets of physicians or hospitals and parents already have a legal duty to care for their child independent of any new fetal right.

Therefore, I recommend the introduction of only the first two of the proposed fetal rights described above, both limited in ways required by the serious objections to any widespread expansion of fetal rights holding against the mother. This would allow United States law to change gradually and with caution as advances in medical science and new medical treatments call for legal reform with a minimal risk of imposing harsh or unjust burdens on pregnant women. At least this would be true if at the same time our social institutions are modified to enable pregnant women to fulfill the legal duties these fetal rights would impose upon them.

CHAPTER 8

THE SCOPE OF THE RIGHT TO PROCREATIONAL AUTONOMY

It is clear that there is some sort of a constitutional right to procreate under United States law. But there is considerable uncertainty about the precise definition of this right today and great controversy about what its scope ought to be in the future. This essay will address both of these issues. It will not, however, consider who does or ought to possess this right, only married persons or perhaps unmarried individuals, lesbian couples, or others. My strategy will be to investigate first the constitutional right to procreate of married persons and then in another chapter to discuss the categories of potential right-holders.

8.1. References to the Right

The right to procreate is not, of course, enumerated in the Constitution itself, but it has been recognized by the Supreme Court of the United States and by other courts on many occasions. One cannot find in any of the opinions of these courts any precise definition or even extended description of the constitutional right to procreate. The best one can do is to examine the language used to refer to this right and attempt to interpret its legal meaning. This will be an essential first step toward defining the scope of the constitutional right to procreate.

The first case in which the Supreme Court referred to a constitutional right to procreate, at least by implication, was *Meyer v. Nebraska*.[1] The issue before the court was whether a statute that prohibited the teaching in any public or private school of a foreign language to a child who had not passed the eighth grade was unconstitutional because it infringed the liberty guaranteed by the Due Process Clause of the Fourteenth Amendment. The opinion of the court hinged on the following passage:

> While this court has not attempted to define with exactness the liberty thus guaranteed, the term has received much consideration, and some of the included things have been definitely stated. Without doubt, it denotes not merely freedom from bodily restraint, but also the right of the individual to contract, to engage in any of the common occupations of life, to acquire useful knowledge, to marry, establish a home and bring up children, to worship God according to the dictates of his own conscience, and, generally, to enjoy those privileges long recognized at common law as essential to the orderly pursuit of happiness by free men.[2]

[1] *Meyer v. Nebraska*, 262 U.S. 390 (1923).
[2] *Ibid.* at 399.

Presumably, the expression "the right of the individual . . . to marry, establish a home and bring up children" refers implicitly to the constitutional right to procreate. This is because the structure of this expression indicates that marriage and bringing up children are tied together within a single right and the traditional common-law link between marriage and raising a child was procreation.

What can one learn about the scope of the right to procreate from this reference to it? First, originally the Supreme Court recognized the right to procreate within the context of the right to marry and to raise one's children. Whether this implies that the right to procreate is conditional on marriage remains to be seen. Second, if the right to establish a home is essentially connected to both marriage and child-rearing, then it presumably includes procreation, that is conceiving and bearing children. This is partially confirmed by the language of the Supreme Court in a subsequent citation of this passage. "The rights to conceive and to raise one's children have been deemed 'essential,' Meyer *v.* Nebraska, 262 US 390, 399 . . . ."[3] Third, procreation is seen as essentially tied to bringing up one's children. Does this imply that child-rearing should be taken to lie within procreation? Some commentators suggest that this is so. "Claims of procreative freedom logically extend to every aspect of reproduction: conception, gestation and labor, and childrearing."[4] And

> We can think of procreation as involving three main elements, one of which is *begetting*—producing offspring that are genetically one's own . . . . The second element is *gestating*, and the third is *rearing* children. Of course, procreators need not participate in all three components.[5]

However, this seems incompatible with the precise language of the Supreme Court. The crucial passage in *Meyer* refers to the right "to marry, establish a home and bring up children." The language used suggests that establishing a home, something that includes procreation, is different from bringing up children. This is confirmed by the citation in Stanley that refers to "the rights [in the plural] to conceive and to raise one's children." Of course, these preliminary conclusions may need to be modified in the light of references to the right to procreate in subsequent cases.

The leading case is clearly *Skinner v. Oklahoma*.[6] At issue was whether the Oklahoma Habitual Criminal Sterilization Act was unconstitutional by reason of the Fourteenth Amendment. Delivering the opinion of the Supreme Court, Justice Douglas began by asserting "This case touches a sensitive and important area of human rights. Oklahoma deprives certain individuals of a right which is basic to the perpetuation of a race—the right to have offspring."[7] A later passage identifies at greater length the right denied by the Oklahoma statute.

> We are dealing here with legislation which involves one of the basic civil rights of man. Marriage and procreation are fundamental to the very existence and survival of the race. The power to sterilize, if exercised, may have subtle,

[3] *Stanley v. Illinois*, 405 U.S. 645 at 651 (1972).
[4] J. A. Robertson, "Procreative Liberty and the Control or Conception, Pregnancy, and Childbirth," *Virginia Law Review* 69, 408.
[5] C. Strong, *Ethics in Reproductive and Perinatal Medicine: A New Framework* (1977), p. 13.
[6] *Skinner v. Oklahoma*, 316 U.S. 535 (1941).
[7] *Ibid.* at 536.

> far-reaching and devastating effects. In evil or reckless hands it can cause races or types which are inimical to the dominant group to wither and disappear. There is no redemption for the individual whom the law touches. Any experiment which the State conducts is to his irreparable injury. He is forever deprived of a basic liberty.[8]

Here is an explicit, but enigmatic, assertion of the constitutional right to procreate, that is to conceive offspring.

The language of the Supreme Court equates the basic civil right at issue with "a basic liberty." Presumably, then, the defining core of the right to procreate is a legal liberty to procreate, to act so as to conceive and produce offspring. What makes it a constitutional right, however, is that this liberty is protected by the Fourteenth Amendment. Hence, an associated element in the right to procreate is a legal immunity against state or federal legislation that would without some overriding justification limit or deny the liberty to procreate.

In two important respects, however, Skinner is unilluminating. First, it does nothing to indicate the boundaries of the core liberty that defines the right to procreate. Surely it does not embrace the liberty to procreate by forcible rape. Whether it includes the liberty to conceive and produce offspring by consensual sex outside of marriage remains a mystery. Nor does it specify which reproductive technologies, if any, may be used in procreation. Second, it does not indicate whether the constitutional right to procreate is an option right, a right either to procreate or not to procreate as the right-holder chooses. Still, it is hard to imagine that even within marriage the Supreme Court intended to recognize a legal duty of either spouse to procreate against her will. If not, then it presumed that the right to procreate is a right to choose whether or not to procreate.

This presumption appears to become explicit in *Eisenstadt v. Baird.*[9] Central to the reasoning of the court is the following passage:

> If the right to privacy means anything, it is the right of the *individual*, married or single, to be free from unwarranted governmental intrusion into matters so fundamentally affecting a person as the decision whether to bear or beget a child.[10]

To be sure, at issue in this case was the constitutionality of a Massachusetts statute that prohibited giving or furnishing contraceptives to any single person except to prevent disease. Hence, it requires an appeal only to the constitutional right *not* to procreate first recognized in *Griswold*. But there is nothing in the language of this passage to limit its meaning to this negative liberty-right. Quite the contrary, it emphasizes "the decision" whether to bear or beget a child. And in support of its application of the constitutional right to privacy, the Court cites three precedents including Skinner, the leading case for the positive liberty-right to procreate.

Moreover, the presumption that the constitutional right to privacy protects procreative decisions is reaffirmed in subsequent judicial reasoning. Thus, the opinion of the Supreme Court in *Roe v. Wade* hinges on this passage:

---

[8] *Ibid.* at 541.
[9] *Eisenstadt v. Baird*, 405 U.S. 438 (1972).
[10] *Ibid.* at 453, italics in original.

> This right to privacy, whether it be founded in the Fourteenth Amendment's concept of personal liberty and restrictions upon state action, as we feel it is, or, as the District Court determined, in the Ninth Amendment's reservation of rights to the people, is broad enough to encompass a woman's decision whether or not to terminate her pregnancy.[11]

And soon afterward, the Supreme Court in *Cleveland Board of Education v. LaFleur* asserted "By acting to penalize the pregnant teacher for deciding to bear a child, overly restrictive maternity leave regulations can constitute a heavy burden on the exercise of these protected freedoms."[12]

Subsequently, after citing a series of cases in which it had recognized that certain areas or zones of privacy are protected by the Fourteenth Amendment, the Supreme Court in *Carey v. Population Services*[13] affirmed that "The decision whether or not to beget or bear a child is at the very heart of this cluster of constitutionally protected choices." Clearly the Court holds that the constitutional right to privacy protects choices, including the decision whether or not to procreate. Finally, in *Davis v. Davis*,[14] the Supreme Court of Tennessee cites both *Skinner* and *Eisenstadt* just before asserting that "a right to procreational autonomy is inherent in our most basic concepts of liberty . . . ."[15] It goes on to explain that

> For the purposes of this litigation it is sufficient to note that, whatever its ultimate constitutional boundaries, the right of procreational autonomy is composed of two rights of equal significance—the right to procreate and the right to avoid procreation.[16]

Although this case is a binding precedent only for Tennessee law, it is grounded solidly upon previous decisions of the United States Supreme Court and is frequently cited in the legal literature. There is, then, substantial evidence to show that what is usually called "the constitutional right to procreate" is an option right, a right either to procreate or not procreate as the right-holder chooses. More precisely, it is a rights-package, a set of rights concerning the same activity or subject matter, consisting of a liberty-right to procreate and a liberty-right to avoid procreation. In order to avoid confusing the rights-package with its positive constituent right, I will henceforth follow the example of *Davis* and refer to the former as the right to procreational autonomy and the latter as the right to procreate.

Davis v. Davis also begins to answer another question about the scope of the constitutional right to procreate, whether the liberty to procreate includes the liberty to use the new reproductive technologies.

> This appeal presents a question of first impression, involving the disposition of the cryogenically-preserved product of *in vitro* fertilization (IVF), commonly referred to in the popular press and the legal journals as "frozen embryos."[17]

---

[11] *Roe v. Wade*, 410 U.S. 113 (1973) at 153.

[12] *Cleveland Board of Education v. LaFleur*, 414 U.S. 632 (1974) at 640.

[13] *Carey v. Population* Services, 431 U.S. 678 (1977) at 684–685.

[14] *Davis v. Davis*, 842S.W.2nd 588 (Tenn. 1992).

[15] *Ibid.* at 601.

[16] *Ibid.*

[17] *Ibid.* at 589, italics in original.

The case arose because Junior Lewis Davis and Mary Sue Davis were able to agree on all the terms of their divorce except who was to have custody of their frozen embryos. The Supreme Court of Tennessee decided this unresolved issue by applying the constitutional right to procreate. In doing so, it did not suggest that the positive liberty to procreate was limited to exclude the use of even the most complex technological means of creating offspring. Quite the contrary, it recognized the relevance of both the liberty to procreate and the liberty to avoid procreation in its reasoning. "The equivalence of and inherent tension between these two interests are nowhere more evident than in the context of in vitro fertilization."[18] The court went on to resolve the dispute by balancing these conflicting interests, that is by weighing the burdens that would be imposed upon Junior Davis by granting his former wife's liberty-right to procreate against the burdens that would be imposed upon Mary Sue Davis by granting her former husband's liberty-right to avoid procreation and decided in his favor.[19] If this precedent is followed by other state courts, then the boundaries of the constitutional liberty to use reproductive technologies will be drawn by balancing the interests of the parties given the specific facts of each case.

What have we learned about the constitutional right to procreate by our examination of references to it in the most important court cases? (1) The right to procreational autonomy is the right to make and act on procreational choice and is a rights-package consisting of the right to procreate and the right to avoid procreation. These do not constitute a single bilateral liberty-right to procreate or not procreate because the content of the positive right to procreate is defined independently of and differently from the content of the negative right not to procreate. In particular, a woman's right to use contraceptives or to have an abortion are very definite and almost unlimited, while her right to use new reproductive technologies has almost no definite boundaries and may well be significantly limited. (2) As components of the right to procreational autonomy, each of its constituent rights is an option right. Thus, the right to procreate is the right to procreate or not procreate as the right-holder chooses, and the right to avoid procreation is the right to decide whether to avoid procreating or not to avoid procreating. (3) Although the constitutional right to procreational autonomy consists of a pair of liberty-rights, its scope is narrower than the full legal procreative liberties of its possessor. For example, one has the legal liberty to procreate or not procreate at whatever time and in whatever manner the law does not prohibit. But only those areas of this liberty that are protected by the constitutional immunity against prohibition or restraint fall within the boundaries of the constitutional right to procreate. (4) The acts of procreation protected by the constitutional right include and are limited to acts of conceiving or not conceiving, gestating or not gestating, and giving birth to or not giving birth to offspring. Although the limits on permissible ways of performing these actions have not been spelled out in detail, they clearly do not exclude the use of many of the means provided by modern medicine.

Examining the ways in which the courts have described the constitutional right to procreational autonomy is only a first step toward defining its scope because these references leave many legal questions unresolved. For example, could states prohibit some means of procreating such as cloning or the use of surrogate mothers? Precisely how do

[18] *Ibid.*, italics in original.
[19] *Ibid.* at 603–604.

the procreative rights of males differ from those of females? Is the liberty to procreate conditional upon one's ability to rear one's offspring? Apparently the nonvoluntary sterilization of very "feeble-minded" females is constitutionally permissible,[20] but would it be permissible to require women receiving welfare benefits to submit to the Norplant system of contraception? Does the right to bear children imply the liberty of bringing an unlimited number of children into existence? And does the right to procreate necessarily exclude the right of a severely defective child to sue one or both parents for wrongful life? One cannot, of course, answer such questions with any confidence before the courts have considered and ruled on the legal issues they present. But the courts could and presumably should decide these issues on the basis of the grounds of the right to procreate. The next step, then, in defining the scope of the constitutional right to procreate must be to examine the authoritative legal sources from which it is derived.

8.2. Grounds of the Constitutional Right

Almost all of the judicial opinions that recognize and interpret the constitutional right to procreational autonomy ground it upon the Due Process Clause of the Fourteenth Amendment, "nor shall any state deprive any person of life, liberty or property, without due process of law." Presumably the Due Process Clause of the Fifth Amendment similarly protects the right to procreational autonomy against denials by the federal government. But precisely how do these provisions of the United States Constitution imply any such right? They each recognize and protect the right to liberty, and the courts have held that the right to procreational autonomy is included in this fundamental constitutional right. However, the reasoning of the Supreme Court has followed two very different paths.

The first of these is exemplified most clearly by the following passage in *Meyer v. Nebraska*.[21]

> While this court has not attempted to define with exactness the liberty thus guaranteed, the term has received much consideration, and some of the included things have been definitely stated. Without doubt, it denotes not merely freedom from bodily restraint, but also the right of the individual to contract, to engage in any of the common occupations of life, to acquire useful knowledge, to marry, establish a home and bring up children, to worship God according to the dictates of his own conscience, and, generally, to enjoy those privileges long recognized at common law as essential to the orderly pursuit of happiness by free men.[22]

Here the right to establish a home, presumably including procreation, is held to be included in the liberty protected by the Due Process Clause because it is implicit in the common-law right to marry.

---

[20] See, for example, In *Re Grady*, N.J., 426 A.2nd 467.
[21] 262 U.S. 390 (1923).
[22] At 399.

William Blackstone's *Commentaries on the Laws of England* indicates that the right to procreate is in fact implicit in the traditional common-law right to marry.[23] In his complex classification of rights, he distinguishes between absolute and relative rights.

> The rights of persons considered in their natural capacities are also of two sorts, absolute, and relative. Absolute, which are such as appertain and belong to particular men, merely as individuals or single persons: relative, which are incident to them as members of society, and standing in relations to each other.[24]

Marriage, of course, constitutes one of the basic private relations between persons.

That the right to procreate is implicit in the common-law right to marry becomes clear when one examines Chapter 14 of Book I, *Of the Rights of Persons*, in Blackstone's *Commentaries*.

> The three great relations in private life are, 1. That of *master and servant*; which is founded in convenience, whereby a man is directed to call in the assistance of others, when his own skill and labour will not be sufficient to answer the cares incumbent upon him. 2. That of *husband and wife*; which is founded in nature, but modified by civil society: the one directing man to continue and multiply his species, the other prescribing the manner in which that natural impulse must be confined and regulated. 3. That of *parent and child*, which is consequential to that of marriage, being it's principal end and design: and it is by virtue of this relation that infants are protected, maintained, and educated.[25]

The right to procreate is an essential presupposition of the English law of marriage both because the relation of husband and wife is founded in nature that directs man "to multiply his species" and because the relation of parent and child is the "principal end and design" of marriage.

Moreover, Blackstone's observation that civil society modifies the relation of husband and wife by "prescribing the manner in which that natural impulse must be confined" suggests that the scope of the common-law right to marry determines the scope of the common-law right to procreate. Because sexual intercourse outside of marriage is contrary to the law, the most significant limits on permissible procreation are those disabilities that limit the legal power of any person to enter into a valid marriage contract.

> Now these disabilities are of two sorts: first, such as are canonical, and therefore sufficient by the ecclesiastical laws to avoid the marriage in the spiritual court; but these in our law only make the marriage voidable, and not *ipso facto* void, until sentence of nullity be obtained. Of this nature are pre-contract;

---

[23] This is not a pure common law right if the common law is defined as that body of law created by national courts and purporting to be derived from natural reason and ancient usage rather than legislation, for it depends also upon canon law and civil statues. But it is a right of the common law defined as the body of English law adopted and adapted by the United States courts and legislatures. The *Oxford English Dictionary* recognizes some such distinction.

[24] W. Blackstone, *Commentaries on the Laws of England*, Vol. I, p. 119.

[25] *Ibid.*, p. 410, italics in original.

> consanguinity, or relation by blood; and affinity, or relation by marriage; and some particular corporal infirmities.[26]

and

> The other sort of disabilities are those which are created, or at least enforced, by the municipal laws. And, though some of them may be grounded on natural law, yet they are regarded by the laws of the land, not so much in the light of any moral offence, as on account of the civil inconveniences they draw after them. These civil disabilities make the contract void *ab initio*, and not merely voidable . . . .[27]

These are a prior marriage or having another husband or wife living, lacking the minimum age required to marry, the absence of parental consent when one has not reached the age of consent, the want of reason, and the failure to contract in due form of law.

Now United States courts could use these disabilities, or other limits on the right to marry developed later in Anglo–American common law, to determine the scope of the constitutional right to procreate. Justice Harlan suggests something like this in his influential dissenting opinion in *Poe v. Ullman*.

> The laws regarding marriage which provide both when the sexual powers may be used and the legal and societal context in which children are born and brought up, as well as laws forbidding adultery, fornication and homosexual practices which express the negative of the proposition, confining sexuality to lawful marriage, form a pattern so deeply pressed into the substance of our social life that any Constitutional doctrine in this area must build upon that basis.[28]

Would this line of reasoning be legally valid and useful in resolving the legal issues likely to arise in our courts?

Its legal validity is probably established by the legal practice of our courts that have consistently cited *Meyer v. Nebraska* as an authoritative source of constitutional law. I doubt, however, that any appeal to the traditional limits on the common-law right to marry would be very useful in deciding cases that hinge on the precise scope of the constitutionally protected liberty to procreate or not procreate. For one thing, they seem irrelevant to most of the legal issues I listed at the end of the previous section. Notice that Harlan does not use the limits he describes to justify his conclusion that the Connecticut antibirth control laws are unconstitutional; he includes it in his opinion to reassure those who might consider his reasoning dangerous. For another thing, it would overturn too many decisions of the Supreme Court by limiting the right to procreate to married individuals. But surely this was not the intention of the Court in *Skinner v. Oklahoma*, the leading case recognizing that right. No one would imagine that a revised Oklahoma statute authorizing the sterilization of only unmarried persons convicted of three felonies would be upheld by the that Court. And it is clearly inconsistent with the frequently cited statement in *Eisenstadt v. Baird*.

---

[26] Blackstone, *op. cit.*, p. 422, italics in original.
[27] *Ibid.*, p. 423, italics in original.
[28] *Poe v. Ullman*, 367 U.S. 497 (1961) at 546.

> If the right of privacy means anything, it is the right of the *individual*, married or single, to be free from unwarranted governmental intrusion into matters so fundamentally affecting a person as the decision whether to bear or beget a child.[29]

Although this case involved the negative right not to procreate, the reasoning of the Supreme Court applies equally to the positive right to procreate, to act on the decision to bear or beget a child.

Nevertheless, there is another way in which United States courts might use the common-law right to marry to determine the scope of the constitutionally protected liberty to procreate. They could appeal to the purposes ascribed to the right to marry in the common law. Blackstone attributes three purposes to this right. The English law regulating the relation of husband and wife permits "man to continue and multiply his species;" results in the relation of parent and child by virtue of which "infants are protected, maintained, and educated;" and rules out the "civil inconvenience" of unlawful marriage.[30] The first two purposes would enable our courts to infer a right to procreate from the right to marry. However, courts could justify limiting the right to procreate in order to protect children or prevent serious civil inconvenience. Quite possibly an examination of later common law would reveal other purposes ascribed to the right to marry that could also be used to define the scope of the constitutional right to procreate. I must reserve judgment on whether this would be a useful method of interpreting constitutional law because I cannot find cases in which it has been attempted by our courts. Rather than follow the lead of *Meyer v. Nebraska* and ground the constitutional right to procreate primarily on the common-law right to marry, most of the later decisions have appealed to the constitutional right to privacy.

The right to privacy is not, of course, mentioned in the text of the United States Constitution. The first case in which the Supreme Court recognized a constitutional right to privacy is *Griswold v. Connecticut*.[31] Unfortunately, the justification for recognizing any such right is far from clear. Although a safe majority of six justices joined in the judgment in that case, only a minority of four joined in the opinion of the Court. In fact, one finds three very different lines of reasoning in the three most influential opinions in *Griswold*. Because each of these suggests a very different method of determining the scope of the constitutional right to procreational autonomy, one needs to assess each in turn.

Justice Douglas delivered the opinion of the Court. The logic of his reasoning is clear. First, he reviewed several prior decisions in which the Supreme Court had derived rights not mentioned in the Constitution from the rights specified in the Bill of Rights. For example, in *Pierce v. Society of Sisters* the right to educate one's children as one chooses was inferred from the First Amendment, and in *Meyer v. Nebraska* the right to study the German language in a private school was similarly grounded on the First Amendment.[32] This right to the freedom of speech or of the press was subsequently construed to include not only the right to utter or print, but also "the right to distribute, the right to receive and the right to read (Martin v Struthers, 319 US 141, 143 . . .)."[33] He reaffirmed the principles

---

[29] *Eisenstadt v. Baird*, 405 U.S. 438 at 453, italics in original.
[30] Blackstone, *op. cit.*, pp. 410, 421.
[31] *Griswold v. Connecticut*, 381 U.S. 479 (1965).
[32] *Ibid.* at 482.
[33] *Ibid.*

of these cases because "Without those peripheral rights the specific rights [mentioned in the Bill of Rights] would be less secure."[34]

Second, he argued that several of these peripheral rights constitute a constitutional right to privacy.

> The foregoing cases suggest that specific guarantees in the Bill of Rights have penumbras, formed by emanations from those guarantees that help give them life and substance. See Poe v Ullman, 367 US 497 . . . (dissenting opinion). Various guarantees create zones of privacy.[35]

Thus, in *NAACP v. Alabama* the disclosure of membership lists was held invalid as entailing the likelihood of a substantial restraint upon the First Amendment right to freedom of association.[36] The Third Amendment's prohibition against the quartering of soldiers in any house in time of peace without the owner's consent is another facet of that privacy. Similarly, the Fourth Amendment affirms the right of people to be secure in their persons, houses, papers, and effects against unreasonable searches and seizures. And the Fifth Amendment's Self-Incrimination Clause "enables the citizen to create a zone of privacy which the government may not force him to surrender to his detriment."[37] These zones of privacy may be referred to collectively as a "right to privacy."[38]

Justice Douglas quoted, without comment or explanation, the Ninth Amendment: "The enumeration in the Constitution, of certain rights, shall not be construed to deny or disparage others retained by the people."[39] He was not here taking the next step towards his conclusion, but responding to critics who would object to his reasoning on the ground that neither the peripheral rights to which he refers nor the right to privacy to which he appeals are mentioned in the text of the United States Constitution.

The third step in his reasoning was to argue that the Connecticut statute prohibiting the use of contraceptives invades the constitutional right to privacy. He did this, not by relying upon any definition of the content of this right to privacy, but by referring back to the Bill of Rights from which he has inferred it.

> Would we allow the police to *search* the sacred precincts of marital bedrooms for telltale signs of the use of contraceptives? The very idea is repulsive to the notions of privacy surrounding the marriage relationship. We deal with a right of privacy older than the Bill of Rights—older than our political parties, older than our school system. Marriage is a coming together for better or worse, hopefully enduring, and intimate to the degree of being sacred. It is an *association* that promotes a way of life, not causes; a harmony of living, not political faiths; a bilateral loyalty, not commercial or social projects. Yet it is an association for as noble a purpose as any involved in our prior decisions.[40]

---

[34] *Ibid.* at 483.
[35] *Ibid.* at 484.
[36] *Ibid.* at 483.
[37] *Ibid.* at 484.
[38] *Ibid.* at 485.
[39] *Ibid.* at 484.
[40] *Ibid.* at 485–486, italics added.

Thus, Douglas was appealing to the Fourth Amendment right against unreasonable searches and seizures and the peripheral constitutional right to freedom of association to show that, at least for married persons, the right to privacy includes the liberty to use contraceptives, one portion of the scope of the right to avoid procreation. Presumably, the same sort of reasoning could be used to justify other conclusions about the scope of both the constitutional right to procreate and the right to avoid procreation.

Could the courts use this method of reasoning to establish the precise scope of the right to procreational autonomy? I doubt that they could do so convincingly. Douglas relegated the right to privacy to a secondary role in his reasoning. Thus, he attempted to justify his conclusion primarily by appealing to rights explicitly mentioned in the Bill of Rights. To be sure, it is the privacy of the marital bedroom that might render its search unreasonable. And he could, although he did not, argue that any prohibition against contraception violates the right to the freedom of association because it invades the privacy necessary to the free association between husband and wife. But he never even attempted to define the precise content of the right to privacy and to explain how the Connecticut statute invades it. Hence, he must rely primarily upon the text of the Bill of Rights. But the relevance of the very general wording of these Amendments to any of the highly specific legal issues concerning procreation is tenuous at best. As Justice Stewart remarked in his dissenting opinion:

> In the course of its opinion the Court refers to no less than six Amendments to the Constitution: the First, the Third, the Fourth, the Fifth, the Ninth, and the Fourteenth. But the Court does not say which of these Amendments, if any, it thinks is infringed by this Connecticut law.[41]

He insinuated, and I am inclined to agree, that Douglas left this out of the opinion of the Court because he was unable to explain precisely how any of the rights mentioned in these Amendments applies to the issue before the Court. Asserting that a constitutional right has a penumbra of undefined scope does little or nothing to provide a logical derivation of a right to privacy applicable to the case before the court.

The concurring opinion of Justice Goldberg suggests, although it does not exemplify, a very different method of determining the scope of the constitutional right to procreational autonomy. Goldberg joined in the opinion of the Court but wrote a separate opinion to explain what it left unexplained.

> In reaching the conclusion that the right of marital privacy is protected as being within the protected penumbra of specific guarantees of the Bill of rights, the Court refers to the Ninth Amendment, ante, at 515. I add these words to emphasize the relevance of that Amendment to the Court's holding.[42]

Thus, his intention was to develop further, not to replace, the method of reasoning used by Douglas.

Nevertheless, his opinion has been misread as resting solely upon the Ninth Amendment. Let us see how his words might suggest this. Goldberg began by explaining the purpose of the Ninth Amendment.

---

[41] *Ibid.* at 527–528.
[42] *Ibid.* at 487.

> The language and history of the Ninth Amendment reveal that the Framers of the Constitution believed that there are additional fundamental rights, protected from governmental infringement, which exist alongside those fundamental rights specifically mentioned in the first eight constitutional amendments.[43]

Hence, the courts are required to recognize these additional fundamental rights as constitutional rights by virtue of this Amendment.

He then argued at length that the right to privacy is one of these additional fundamental rights.

> In determining which rights are fundamental, judges are not left at large to decide cases in light of their personal and private notions. Rather, they must look to the "traditions and [collective] conscience of our people" to determine whether a principle is "so rooted [there]... as to be ranked as fundamental."... The inquiry is whether a right involved "is of such a character that it cannot be denied without violating those 'fundamental principles of liberty and justice which lie at the base of all our civil and political institutions'...."... I agree fully with the Court that, applying these tests, the right of privacy is a fundamental personal right, emanating from the totality of the constitutional scheme under which we live."[44]

He here assumed that the opinion of the Court has applied these tests by appealing to the Bill of Rights and emanations from it.

Finally, he argued that the contested statute prohibiting the use of contraceptives is unconstitutional because it infringes the constitutional right to privacy.

> Although the Connecticut birth-control law obviously encroaches upon a fundamental personal liberty, the State does not show that the law serves any "subordinating [state] interest which is compelling" or that it is "necessary... to the accomplishment of a permissible state policy."[45]

Accordingly, he joined in the judgment of the Court "that petitioners' convictions must therefore be reversed."[46]

Although Goldberg presupposed that Douglas has established the constitutional right to privacy by reasoning from the Bill of Rights, others might use the tests for fundamental rights he proposed to reason directly from the Ninth Amendment to this right. Indeed, the United States District Court purports to be using this method of reasoning in its opinion in *Roe v. Wade*.[47]

> On the merits, plaintiffs argue as their principal contention that the Texas Abortion Laws must be declared unconstitutional because they deprive single

---

[43] *Ibid.* at 488.
[44] *Ibid.* at 493–494.
[45] *Ibid.* at 497–498.
[46] *Ibid.* at 499.
[47] *Roe v. Wade*, 314 F.Supp. 1217 (1970).

> women and married couples of their right, secured by the Ninth Amendment, to choose whether to have children. We agree.[48]

Apparently, then, the District Court intended to base its decision squarely upon the Ninth Amendment.

It began its reasoning regarding the legal issue presented by this case precisely as one would expect:

> The essence of the interest sought to be protected here is the right of choice over events which, by their character and consequences, bear in a fundamental manner on the privacy of individuals. The manner in which such interests are secured by the Ninth Amendment is illustrated by the concurring opinion of Mr. Justice Goldberg in Griswold v Connecticut.[49]

But how are privacy interests secured by the Ninth Amendment? Presumably by the fact that the right to privacy is a fundamental right given constitutional status by that Amendment. And in footnote 9 on the same page, the District Court refered to Goldberg's tests quoted above. Accordingly, one would expect the Court next to use these tests to establish the constitutional right to privacy and then to argue that the individual's right to choose whether to have children (the right to procreational autonomy) falls within the scope of this right. If so, its decision that the Texas Abortion Laws are unconstitutional would be determining one portion of the scope of the right not to procreate by a method of reasoning very different from that used in the opinion of the Supreme Court in *Griswold*.

In fact, probably because of the inadequacy of these tests to identify and define the fundamental rights recognized by the Ninth Amendment, it did not do so. Instead it argued that "Relative sanctuaries for such 'fundamental interests' have been established for the family, the married couple, and the individual." The three footnotes to this crucial sentence refer for the most part to *Pierce v. Society of Sisters*, *Meyer v. Nebraska*, *Prince v. Commonwealth of Massachusetts*, *Loving v. Virginia*, *Skinner v. Oklahoma*, and *Stanley v. Georgia*. Now all of these cases were decided on the basis of the Fourteenth Amendment, either its Due Process Clause or its Equal Protection Clause, not by any direct reasoning from the Ninth Amendment. In this regard, the third conclusion of law drawn by the District Court is illuminating. "The fundamental right of single women and married persons to choose whether to have children is protected by the Ninth Amendment, *through the Fourteenth Amendment*."[50] This suggests that a more promising method of reasoning about the scope of the constitutional right to procreational autonomy would rest primarily, not upon the Ninth Amendment, but upon the Fourteenth.

The opinion of Justice Harlan in *Griswold* clearly presupposes this third method of reasoning. Although he concurred in the judgment of reversal in that case, he found himself unable to join in the opinion of the Court.

> In my view, the proper constitutional inquiry in this case is whether this Connecticut statute infringes the Due Process Clause of the Fourteenth Amendment

[48] *Ibid.* at 1221.
[49] *Ibid.*
[50] *Ibid.* at 1225, italics added.

> because the enactment violates basic values "implicit in the concept of ordered liberty," Palko v Connecticut.... For reasons stated at length in my dissenting opinion in Poe v Ullman, supra, I believe that it does. While the relevant inquiry may be aided by resort to one or more of the provisions of the Bill of Rights, it is not dependent on them or any of their radiations. The Due Process Clause of the Fourteenth Amendment stands, in my opinion, on its own bottom.[51]

Thus Harlan rejected the reasoning of Douglas and Goldberg and proposes instead to ground the constitutional right to privacy on the Due Process Clause.

One must look to his dissenting opinion in *Poe v. Ullman* to find the method of reasoning he advocates. There he explains at some length how he believes that one should interpret the Due Process Clause in terms of the concept of ordered liberty.

> Due process has not been reduced to any formula; its content cannot be determined by reference to any code. The best that can be said is that through the course of this Court's decisions it has represented the balance which our Nation, built upon postulates of respect for the liberty of the individual, has struck between that liberty and the demands of organized society. If the supplying of content to this Constitutional concept has of necessity been a rational process, it certainly has not been one where judges have felt free to roam where unguided speculation might take them. The balance of which I speak is the balance struck by this country, having regard to what history teaches are the traditions from which it developed as well as the traditions from which it broke.[52]

Therefore, "Each new claim to Constitutional protection must be considered against a background of Constitutional purposes, as they have been rationally perceived and historically developed."[53]

Although the privacy of the home receives explicit constitutional protection in only two Amendments, Harlan asserted that this Court has held and today confirms "that the concept of 'privacy' embodied in the Fourth Amendment is part of the 'ordered liberty' assured against state action by the Fourteenth Amendment."[54] He found the most comprehensive statement of the right to privacy implicit in the concept of ordered liberty in a previous opinion of Justice Brandeis.

> "The protection guaranteed by the [Fourth and Fifth] Amendments is much broader in scope. The makers of our Constitution undertook to secure conditions favorable to the pursuit of happiness. They recognized the significance of man's spiritual nature, of his feelings and of his intellect. They knew that only a part of the pain, pleasure and satisfactions of life are to be found in material things. They sought to protect Americans in their beliefs, their thoughts, their emotions and their sensations. They conferred, as against the

---

[51] *Griswold* at 500.

[52] *Poe v. Ullman*, 367 U.S. 497 at 542.

[53] *Ibid.* at 544.

[54] *Ibid.* at 549.

> Government, the right to be let alone—the most comprehensive of rights and the right most valued by civilized man. To protect that rights, every unjustifiable intrusion by the Government upon the privacy of the individual, whatever the means employed, must be deemed a violation of the Fourth Amendment...."[55]

Thus, one of the rational purposes of the Constitution, including the Due Process Clause, is to protect the privacy of the individual against state intrusions.

To determine the boundaries of this constitutional right to privacy, one must see how this rational purpose has been historically developed in the traditions of our nation.

> Adultery, homosexuality and the like are sexual intimacies which the State forbids altogether, but the intimacy of husband and wife is necessarily an essential and accepted feature of the institution of marriage, an institution which the State not only must allow, but which always and in every age it has fostered and protected. It is one thing when the State exerts its power either to forbid extra-marital sexuality altogether, or to say who may marry, but it is quite another when, having acknowledged a marriage and the intimacies inherent in it, it undertakes to regulate by means of the criminal law the details of that intimacy.[56]

Clearly, then, although the constitutional right to privacy is not absolute, it is broad enough to protect marital privacy. Therefore, the Connecticut statute prohibiting the use of contraceptives by married couples ought to be declared unconstitutional.

Of the three methods of reasoning about the constitutional right to procreational autonomy one finds in the opinions of *Griswold*, the one proposed by Justice Harlan is by far the most promising because it provides the resources to infer specific applications from the constitutional right to privacy without preventing re-interpretation called for by new cases. By building upon the legal and social traditions as historically developed in our society, it provides a wealth of specific detail unavailable to either Douglas or Goldberg. At the same time, by appealing to the rational purposes of the Constitution and of our social institutions, such as marriage, it permits the courts to recognize why those traditions have changed historically and even to develop our legal traditions further to enable them to achieve those purposes under changed circumstances. At the same time, the richness of these resources to which the courts can and should appeal makes it difficult if not impossible for legal scholars to predict with any precision how the Supreme Court will draw the boundaries of the right to procreate in the future. Ronald Dworkin has suggested that the constitutional right to procreational autonomy is also grounded upon the First Amendment right to freedom of religion.[57] However, I will not discuss his reasoning because I can find almost no decisions of the Supreme Court to enable one to assess its strength. It might, however, be useful for a philosopher of law to reflect upon what the scope of the constitutional right to procreate ought to be.

---

[55] *Olmstead v. United States*, 277 U.S. 438 (1928) at 478.
[56] *Ibid.* at 553.
[57] R. Dworkin, "The Concept of Unenumerated Rights," *University of Chicago Law Review* 59 (1992), 415–426.

8.3. The Human Right to Procreational Autonomy

Other things being equal, the law ought to respect, and if possible secure, our fundamental human rights. *Skinner v. Oklahoma*, the first case in which the Supreme Court explicitly recognized the constitutional right to procreate, may well presuppose this moral principle. The opinion of the Court begins by asserting: "This case touches a sensitive and important area of human rights. Oklahoma deprives certain individuals of a right which is basic to the perpetuation of a race—the right to have offspring." If there really is a human right to procreate, then this is surely relevant to how the courts ought to define the constitutional right to procreate.

Article 16 § 1 of the *Universal Declaration of Human Rights* asserts "Men and women of full age, without any limitation due to race, nationality or religion, have the right to marry and to found a family...." Although this right may be broader than the right to procreate, because one might be said to found a family by adopting one or more children, it presumably includes the right to have offspring. Now the United Nations intended the *Universal Declaration* to be both a basis for the later Covenants that would introduce human rights into international law and an affirmation of those fundamental moral rights that ought to be legally protected. Although some sort of a right to found a family has become established in international law, its definition is even less determinate and more controversial than that of the constitutional right to procreate. Let us, therefore, leave its consideration for lawyers and turn our attention to the most plausible conception of the human right to procreate when this is assumed to be a basic moral right.

The defining core of this human right is presumably the liberty to procreate, where to procreate is to produce offspring. In spite of the development of reproductive technologies, this remains something that no individual human being can do alone; it takes two to procreate. Thus, strictly speaking the core of the human right to procreate is the moral liberty, at least of a married individual human being, to engage in any of those actions that are severally necessary and collectively sufficient for producing offspring. For a male, these are only impregnating a female; for a female, these include conceiving, gestating, and delivering a child. Offhand, I can see no reason why this moral liberty is not broad enough to permit the use of most of the reproductive techniques available in modern medicine, provided these are reliable and safe. Whether the use of a surrogate mother is permissible is another question best left for another occasion.

The general moral liberty to procreate thus understood is an innocent liberty. That is to say that there is no general moral duty not to produce offspring per se. Some might disagree. Those who accept Buddhism or, like Schopenhauer, its basic presuppositions believe that the human condition is by its very nature miserable because to be alive is to have desires and desires are essentially insatiable. More recently some have suggested that overpopulation now threatens the entire human race with misery because the limited resources in our world will soon be inadequate to sustain even a minimally satisfactory life. However, I do not find either consideration sufficient to impose a general moral duty upon all human beings not to produce offspring, nor do those who advocate either of these two pessimistic viewpoints.

Nevertheless, there probably are more specific duties not to procreate that limit the scope of the moral liberty to procreate. One of these is the moral duty not to procreate without the consent of one's sexual partner. It is not merely that there is a duty not to rape another person. There is also a moral duty not to impregnate or conceive by deceiving one's partner, for example by causing one's partner to believe that one is taking contraceptive measures when one is not doing so. This wrongs one's nonconsenting partner because it is a betrayal of her or his trust and may well result in the birth of a child thus imposing burdensome responsibilities upon him or her. Another limiting duty is the duty not to knowingly risk giving birth to a child so seriously defective that she would have a life not worth living.[58] Surely there is a duty not to act now so as to risk causing one's child to suffer so severely in the future. A third limiting duty is probably the duty not to procreate when one knows or ought to know that one is unwilling or unable to rear, or arrange for others to rear, one's child adequately.[59] Rearing includes both caring for the child while she is unable to care for herself and preparing her to live independently when she becomes an adult. One's liberty to procreate is not limited to some definite number of children, but one does have a duty not to produce more children than one can rear adequately.

The precise extent to which these moral duties limit the moral liberty to procreate or not procreate depends on one additional consideration. "Ought" implies "can" not merely because to say that someone morally ought to act in some manner presupposes that they are able to act in this manner, but also because it presupposes that they can do so without excessive sacrifice.[60] Now to refrain from procreating often involves great sacrifice for one or both sexual partners because this often means giving up the creation of a new human life, the experiences of pregnancy and childbirth, the shared experiences of raising a child of one's own and other values that would immeasurably enrich one's life.[61] How great such sacrifices are will depend upon the desires and self-images of the individuals concerned and whether they already have one or more children. But surely they are sometimes great enough to undermine a woman's prima facie duty not to have a child when she might give birth to a moderately defective child or the duty of a husband not to have a child when he is unsure of his ability to provide for the needs of that child adequately. Because the amounts of harm risked and the degrees of sacrifice involved in acting or not acting to procreate vary so greatly with the circumstances, it is impossible even roughly to draw in general terms the boundaries of the moral liberty to procreate. All one can do is to specify, as I have done, the relevant considerations that are applicable to each individual case.

On the other hand, I do not believe that there is any general moral duty to exercise one's liberty to procreate. Even Christians who believe that God commanded Adam and Eve to multiply and propagate their species need not infer that this imposes a moral obligation to reproduce upon every human being. Indeed, many would regard it sinful for unmarried persons to fornicate and beget offspring; and although marriage presumably confers a liberty to procreate, it would impose a duty to do so only if there were some such understanding with one's spouse. Nor can I imagine any other moral duty-imposing reason that would ground a general duty to procreate. Because a moral liberty in the Hohfeldian sense in which I am using the term is by definition simply the absence of any

---

[58] Compare L. M. Purdy, *Reproducing Persons: Issues in Feminist Bioethics* (Ithaca: Cornell University Press, 1996), pp. 45–49 and R. M. Weinberg, "Procreative Justice: A Contractualist Account," *Public Affairs Quarterly* 16 (2002), 420.

[59] See O. O'Neill, Begetting, bearing, and rearing. In O. O'Neill and W. Ruddick (eds.), *Having Children: Philosophical and Legal Reflections on Parenthood* (New York: Oxford University Press, 1979), pp. 25, 29–30.

[60] Wellman, *Real Rights*, pp. 63–65.

[61] See C. Strong, *Essays in Reproductive and Perinatal Medicine* (New Haven & London: Yale University Press, 1997), pp. 18–22.

contrary duty, one has a general moral liberty not to procreate as well as a moral liberty to procreate. Therefore, the defining core of the human right to procreate is a bilateral moral liberty either to engage in any of those actions that normally are severally necessary and collectively sufficient for producing offspring within the limits I have just specified or not to so act. Thus, the right to procreate is an option right, a right to choose whether or not to engage in any action that, given the necessary conditions, normally results in the birth of a child.

In addition to its defining core, any real right consists of associated Hohfeldian positions that, if respected, confer dominion over that core upon the right-holder. For the moral right to procreate, the most relevant of these is the moral claim of the possessor against others that they not interfere with her exercise of her core bilateral liberty. This moral claim holds against the world, including the government and one's sexual partner. Accordingly, it imposes a moral duty upon all second parties not to prevent or coercively hinder one from engaging in procreative activities and, on the other hand, not to force or put pressure upon one to procreate against her will.

The constitutional right to procreational autonomy is a rights-package that includes both a right to procreate and a right to avoid procreation. Is there a human right to procreational autonomy that similarly includes a both procreational rights? It might seem that if the human right to procreate is an option right including the liberty not to engage in acts of procreating, there is no room left for a distinct human right anything like the constitutional right to avoid procreation, but this is not so. It is one thing to refrain from actions that in the normal course of events result in offspring and quite another to take positive actions to prevent the birth of a child. Therefore, there can be a human right to procreational autonomy that is a rights-package consisting of the human right to procreate and the human right to prevent procreation.

The defining core of the human right to prevent procreation is the bilateral moral liberty either to act so as to prevent the procreative process from beginning, as by using contraceptives, or to terminate the procreative process, as by abortion, or to refrain from any such actions. Although the liberty to prevent the procreative process from beginning is probably in general an innocent liberty because there is no general duty to procreate, the general liberty to terminate the procreative process once it has begun is a suspect liberty. This is because there is a general prima facie moral duty not intentionally to destroy a potential human being grounded on the fact that most human beings have a reasonably valuable life together with the moral principle that one ought not to prevent the creation or development of anything of great value. Exceptions to this general duty include at least cases where the child would be born so seriously defective as to have no chance of a worthwhile life, where the parents are unwilling or unable to rear the child adequately or where the birth of the child would impose an undue sacrifice upon one or both parents. Accordingly, the scope of the moral liberty to prevent procreation probably includes the liberty to prevent procreation at least under these three conditions.[62] Thus, the human right

---

[62] Some persons might suggest that there is another limiting condition, the consent of one's sexual partner. Assuming that there is a moral right to privacy, I do not believe that a pregnant female requires the consent of her partner to make having an abortion morally permissible. On the other hand, the male partner ought not to administer medication to cause an abortion by deception rather than by obtaining the consent of or at the request of his sexual partner.

to prevent procreation is an option right to choose whether to prevent procreation under one or more of these specified conditions or not to do so. This core liberty is presumably protected by a moral claim against others not to interfere with its exercise.

Here, then are plausible, but incomplete, definitions of both a human right to procreate and a human right to prevent procreation that together would constitute a human right to procreational autonomy. Whether some such human rights really exist remains undecided. This depends upon whether there are moral reasons sufficient to ground both a moral liberty-right to procreate and a moral liberty-right to prevent procreation of the sorts I have described. Assuming for the sake of the argument that there really are some such rights, ought the scope of the constitutional right to procreational autonomy to match that of the analogous human rights-package? I believe that, far from being morally required, this would be morally wrong. Notice that the scope of the human right to procreational autonomy is relatively narrow. It rules out choosing to procreate when there is a risk of producing a seriously defective child or when the parents are unwilling or unable to rear their child adequately. And it excludes either procreating or preventing procreation without the consent of one's sexual partner. But for the law to prohibit a wife from having an abortion without the consent of her husband or a husband from having a vasectomy without the consent of his wife presumably threatens the human right to privacy of wife or husband by intruding into a private decision.

8.4. The Human Right to Privacy

Article 12 of the *Universal Declaration of Human Rights* asserts that "No one shall be subjected to arbitrary interference with his privacy, family, home or personal correspondence, nor to attacks upon his honour or reputation. Everyone has the right to the protection of the law against such interference or attacks." And Article 8 of the *European Convention for the Protection of Human Rights and Fundamental Freedoms* affirms a "Right to respect for private and family life." Because international courts have not yet defined the content of this right with precision and because the relevance of the international law of human rights to United States law is unclear, I will assume only that these passages may presuppose the existence of a fundamental moral right to privacy.

I believe that the moral human right to privacy is a rights-package consisting of the four human rights to the privacy of one's spaces, information about one, one's experiences and one's decisions. Each of these component rights is defined by a moral claim against others that they not intrude into or invade some private area of one's life. But the way in which another might invade one's private space by breaking into one's home or bursting into one's bedroom without knocking is quite different from the way in which one intrudes into the privacy of information about one by recording one's telephone conversation or publishing news of one's extramarital affairs. Similarly, the way in which someone invades one's experience by playing his radio so loudly that one cannot avoid the noise is very different from the way in which the legal prohibition of abortion would intrude into the decisions of pregnant women. Moreover, the factors that make one's spaces, information about one, one's experiences and one's decisions private are not, as far as I can see, reducible to one common denominator. Therefore, the human right to privacy should be taken to be a rights-package consisting of four distinct privacy rights rather than a single general right to privacy.

The defining core of the human right to the privacy of one's decisions is a moral claim against others that they not intrude into any of one's private decisions. Others intrude into or invade this area of privacy whenever they prevent or hinder one from acting on a private decision. Hence, this core claim is a moral claim against interference with certain kinds of actions. A private decision is one that affects the life of the agent for better or worse and such that no one else has a sufficient reason to prevent or hinder one from acting on it.[63] It is a choice such that how one acts makes a difference to the value of one's own life and is no one else's business.

A moral claim of one party is the logical correlative of a relative duty of some second party. Hence, a moral claim of X against Y consists of a duty of Y together with a power of X to claim performance of this duty.[64] Accordingly, in order to explain the grounds of the moral claim that others not intrude into any of one's private decisions, it is necessary to explain both the grounds of the duty of others not to interfere with one's acting on a private decision and the grounds of one's power to claim performance of this duty.

There is a general prima facie moral duty not to prevent or hinder any human being from acting as she chooses. Any interference with one's decision is experienced as an invasion, an intrusion by an alien and unwelcome force. As such it is experienced as disturbing and frustrating and normally resented to a degree that damages personal relationships. Interferences with one's decisions also disrupt one's activities even if these are only quiet relaxation or taking a walk. Often they prevent or hinder one from pursuing projects of considerable importance to one. Thus, they undermine one's control over one's projects and deprive one of the values one could have achieved from them. Finally, any interference with doing as one chooses expresses disrespect for one's capacity to make rational decisions because it suggests that left to one's own devices one is unable or unwilling to act as one should. Thus, it undermines that self-respect essential to sustain the agent's dispositions to make difficult decisions and to persevere in carrying them out, both capacities required for living well. These harmful consequences of preventing or hindering one from doing as one chooses are moral reasons for others not to interfere in one's decisions. The fact that someone has disregarded these harms and nevertheless interfered with the agent's decision is also a reason for others to react negatively to him because it shows him to be at least inconsiderate and perhaps domineering and willing to spoil projects valued by others, characteristics that threaten sociability.

This duty not to intrude into the decisions of another is, however, only a prima facie duty. It is sometimes, perhaps often, overridden by some stronger duty to intervene. For example, the duty to prevent rape or to protect someone's life or even property justifies one in interfering with some decision of an agent in spite of the fact that one's intrusion will be experienced as distressing and disrespectful and precisely because it will disrupt the wrongful action of that person. But this is never true when someone is acting on a private decision, for by definition a private decision is one such that no one else has a sufficient reason to interfere. And this aspect of private decisions makes the moral duty not to intrude more stringent because it is then impossible for anyone who intrudes to justify his action to the agent whose decision has invaded. Were one able to explain to that person why one has prevented or hindered her action, this would show that one's motive was not hostility

---

[63] Se the following essay for a fuller explanation of a private decision.

[64] C. Wellman, *A Theory of Rights*, pp. 136–146.

or any indifference to her well-being. This would tend to reduce her resentment and, to the degree that she is a reasonable person, reconcile her to having her action disrupted. But the fact that one has no sufficient reason to intrude would tend to increase resentment because it shows the intruder to be self-centered and domineering.

Moreover, the duty not to intrude into a decision of another is even stronger when that person is acting on a personal decision. Here I am using the word "personal" in the sense defined by the *Oxford English Dictionary* as "Of, pertaining to, concerning, or affecting a person as an individual (rather than as a member of a group or of the public, or in a professional capacity etc.)" A decision is personal in this sense when how one ought to act depends on such things as the feelings, goals, value judgments, and projects of the agent, factors that vary greatly from person to person. Because others are usually not in a position to assess these factors accurately, intervention in a personal decision is more likely to be misplaced and harmful than normal, and one who takes upon himself to substitute his judgment for that of the agent shows himself to be arrogant. This does not imply that one is never justified in interfering with a personal decision. Still, the greater chance of causing harm is a special reason not to intrude into private decisions and the arrogance an intruder displays is an additional reason for others to react negatively should he act contrary to this reason.

Finally, the duty not to intrude into the decisions of another is much stronger when that person is acting on a momentous decision. A momentous decision is one that greatly affects one's life as a whole for better or worse. Examples would be decisions concerning the medical treatment of life-threatening diseases, or whom to marry or which career to pursue. These decisions determine long-term projects and involve commitments that will greatly affect most aspects of one's life for better or worse. Hence, when one intrudes into a momentous decision, the amount of harm one may well cause is magnified and the degree of hostility or at least indifference one displays greater. In these ways, the moral reasons that ground the duty not to intrude are intensified and the strength of the duty they ground correspondingly increased.

But does one have the moral power to claim performance of the duty of others not to intrude into one's private decisions? Yes, because the act of demanding or requesting the duty-bearer to perform this duty together with presenting one's title as claimant, perhaps by pointing out that a private decision is at issue, changes the situation in a way that renders the duty more peremptory. Whenever one violates one's duty not to interfere with a private decision, one intrudes into the personal experience of another and disrupts her activities while at the same time one shows oneself to be at least inconsiderate and perhaps domineering and willing to spoil projects valued by others. But if one interferes with a private decision after the right-holder has requested or demanded that one not do so and pointed to the ground of her claim, then one flagrantly disregards the wishes of another in a way that will be more deeply resented and damage their personal relationship more irreparably. And one will also thereby confirm the fact that one is domineering and show oneself to be hostile to the claimant's projects. Hence, the reasons not to intrude, the harms to the right-holder, will be increased and the reasons for others to respond negatively, the personality traits damaging to sociability, will be revealed to be more undeniable and objectionable.

These moral reasons are, I believe, sufficient to ground a fundamental moral claim against others that they not intrude into any of one's private decisions by preventing or hindering one from acting as one chooses. To be sure, it is true by definition that others never have a sufficient reason to interfere with a private decision. However, one does not have any moral claim against others that they never act without sufficient reason, for many of their decisions will never affect one in any way. One's moral claim against invasions of the privacy of one's decision is grounded on the harms of such intrusions together with the antisocial character traits shown by anyone who does intrude into one's private decisions. This moral claim might well constitute the defining core of a human right to the privacy of one's decisions. Presumably any full moral right to privacy must also include various associated moral positions such an immunity from having others extinguish this core claim by any one-sided action and the liberty to claim performance of the correlative duty. Assuming for the sake of the argument that this is so, the question remains as to how, if at all, this human right to privacy applies to procreational decisions.

Are the decisions of married individuals to procreate or refrain from procreating private? Generally they are private, at least in societies such as ours, because under normal circumstances no one else has a sufficient reason to prevent or hinder one from acting on a decision either to procreate or to refrain from procreating. This does not mean that others have no reason to interfere with such decisions. A wife who will feel fulfilled only if she has a child has a reason to prevent her husband from refraining from sexual intercourse, and the state has a reason to prevent the birth of a child who can be adequately cared for only with public funding. However, in most cases the reasons others have to intrude into decisions to procreate or refrain from procreation are not sufficient to override the reasons against coercive interference. Their insufficiency depends primarily upon the fact that such decisions are typically both personal and momentous. As Dan Brock observes, "Few decisions that people make are more personal than these, in the sense that what is the best choice depends on people's own personal aims and values, or more far-reaching in their impact on people's lives."[65] What makes them personal is that the value of having or not having a child depends upon variables such as how committed one is to one's career, if any, whether one's self-image places motherhood or fatherhood central, and the degree to which one is family oriented. What makes them momentous is that having or not having a child will radically change one's life in a way that will promote or damage many of one's activities and projects for a very long time, in most cases even after the child has ceased to be economically dependent upon one.[66] Hence the harm suffered by the person who makes the wrong choice will almost always outweigh the harm inflicted upon anyone else. Also the personal nature of the decision means that the individual deciding whether to procreate or not is normally in a much better position to make the right choice than anyone else tempted to intrude into the decision. Finally, the fact that interfering to prevent or hinder one from acting on any such decision is coercive, not merely undesired but against the wishes of the person whose decision is obstructed, means that it almost inevitably shows disrespect for the rational agency of that person. When one interferes with a decision as personal and momentous as one concerning procreation, one exhibits a flagrant disrespect that causes a correspondingly great harm. Therefore, under normal circumstances decisions to procreate or refrain from procreating are private.

---

[65] D. W. Brock, Reproductive freedom: Its nature, bases, and limits. In D. Thomasma and J. Monagle (eds.), *Health Care Ethics: Critical Issures for Professionals* (Gaithersbrug MD: Aspen Publishers, 1994), p. 49.

[66] Compare J. A. Robertson, "Liberalism and the Limits of Procreative Liberty: A Response to My Critics," *Washington and Lee Law Review* 52 (1995), 235–236.

There are, as one would expect, exceptions. A husband's decision to procreate by raping his wife is not a private decision. However, this is not because it is a decision to procreate, but because it violates the bodily integrity of the victim and will probably impose upon her the unwelcome burdens of pregnancy, delivery, and subsequently child-care. And the decision to have a child knowing that it will almost certainly be born suffering from a serious genetic defect may not be private because others might be justified in preventing the birth of a child condemned to suffer a miserable life. But decisions such as these are exceptions precisely because they are abnormal and very infrequent. In the case of decisions to refrain from procreation, exceptions are even more unusual.

Are the decisions of married individuals to prevent or not prevent procreation also private? Presumably they are private because the considerations described above apply equally to them. Typically any decision of a married person to prevent or not prevent procreation is both personal and momentous in exactly the same way that any decision to procreate or not procreate is. Hence, any intrusion to prevent or hinder one from acting on any such decision expresses flagrant disrespect for the rational agency of the person whose decision is thus obstructed. Therefore, under normal circumstances the decisions of married individuals either to prevent or not to prevent procreation are private. However, there are exceptions. A husband's decision to prevent procreation by a physical attack upon his pregnant wife to cause a miscarriage is not private, nor in some cases would a wife's decision to refrain from preventing procreation when she knows that her child will be born suffering from one of the very worst genetic defects. Thus, it would seem that decisions to prevent or not prevent procreation are in general, but not always, private just as decisions to procreate or refrain from procreating are.

Still, there does seem to be a difference between contraception and abortion. Presumably a decision to prevent procreation by using contraceptives, especially with the consent of one's spouse, is private because there is no victim when one acts on this decision. One does not violate any right of or do any harm to the child whose procreation one thereby prevents because that child does not yet and never will exist to have any rights or to be harmed. But a human fetus does exist and does seem to be the victim of a pregnant woman's decision to have an abortion. This is not, as some believe, because killing a human fetus violates its human right to life. Rights concern the proper allocation of dominion, freedom, and control, in some potential conflict of wills. Because only moral agents are capable of exercising either freedom or control, only moral agents are possible right-holders. And because the human fetus lacks moral agency, it cannot possess any moral rights, even the fundamental human right to life.[67]

Nevertheless, does not a decision to have an abortion, and thereby prevent the birth of a child who could have lived a healthy happy life, harm the unborn child in the most serious of ways? If so, others might have a sufficient reason to interfere with the decision of the pregnant woman in order to protect her fetus from this great harm. In spite of appearances, I believe that this is not so. Leaving aside the instrumental value of a human life, its consequences for the lives of others, its value for the person whose life it is consists in whether one finds it satisfying or unsatisfactory in the living, in whether one experiences it as being desirable or undesirable. Therefore, only a sentient being, one

[67] C. Wellman, *Real Rights*, pp. 107–113, 137–145.

capable of being conscious or having experiences, could possibly be harmed or benefited. However, isolated experiences, no matter how good or bad they may feel, are never more than trivial. What matters greatly to the value of one's life is the meaning or significance of those experiences that sum up much of one's past and anticipate extended stretches of one's future. Of most importance is one's awareness of success or failure in projects central to one's life. Therefore, only a being having a life, capable of experiencing one's biography as going well or badly, could possibly be seriously harmed or benefited.

What is required in order for one to have a life? Tom Regan gave a highly plausible account.

> An alternative to viewing being-alive as the relevant similarity is what will be termed *the subject-of-a-life criterion*. To be the subject-of-a-life, in the sense in which this expression will be used, involves more than merely being alive and more than merely being conscious. To be the subject-of-a-life is to be an individual whose life is characterized by those features explored in the opening chapters of the present work: that is, individuals are subjects-of-a-life if they have beliefs and desires; perception, memory, and a sense of the future, including their own future; an emotional life together with feelings of pleasure and pain; preference- and welfare-interests; the ability to initiate action in pursuit of their desires and goals; a psychophysical identity over time; and an individual welfare in the sense that their experiential life fares well or ill for them, logically independently of their utility for others and logically independently of their being the object of anyone else's interests.[68]

Although I reject Regan's thesis that this is a sufficient condition for the possession of moral rights, I agree that having a life is what makes one morally considerable in a most important way because it makes one capable of being seriously harmed or benefited by how the actions of a moral agent affect one's life. Because a human fetus is nothing like the subject of a life in this sense, it cannot be seriously harmed by the decision of a pregnant woman to have an abortion.

This does not imply that there is nothing morally wrong with having an abortion. At some point in its development, a human fetus probably becomes capable of feeling pain. If aborting a fetus late in pregnancy really does result in a "silent scream," then the fetus is harmed by the decision of a pregnant woman to have an abortion. This might well be a reason for others to intrude to prevent her from acting on her decision. But the pain suffered by the fetus is not in general a harm sufficient to outweigh the harms that would be imposed on a woman forced to carry her fetus to term and to deliver and care for an unwelcome child.

Also, a woman who decides to have an abortion merely for some trivial reason, such as to remain slim enough to look attractive in her new bathing suit, does seem to be insensitive and uncaring. Compare our feeling that something important has been lost when a talented but troubled youth commits suicide or by the premature death of a woman stricken down by an infection in the prime of her life. We grieve, not just for the survivors, but that the deceased will never enjoy the potential goods that could have been

---

[68] T. Regan, *The Case for Animal Rights* (Berkeley: University of California Press, 1983), p. 243. Italics in the original.

realized in their lives. And one who does not grieve at such a loss is somehow morally defective. Similarly, one who does not regret the loss of a potentially healthy and happy life when a fetus is aborted is insensitive. Although the human fetus does not yet possess the psychological capacities that would make it capable of being harmed or benefited, it does have the potentiality of becoming such a being. Indeed, it already is the human being that will have these capacities if it is born. This is because a human being has a psychophysical identity. A human being is individuated by her living body as much as by her psychological life. And under normal circumstances, a human fetus is a living organism that will become a normal adult human being. An important virtue is a concern for the potential well-being of others, an appreciation of and caring for what others could, given favorable conditions, enjoy or suffer. Conversely, a character trait destructive of sociability is callousness or a lack of appreciation for possible human values. And it does seem that morally one ought to cultivate one's virtues and at least not act in ways that will tend to make one a morally less good person. Thus, if having an abortion sometimes tends to harden the heart of a pregnant woman and to make her less caring about potential human welfare, then this is a reason to which others might appeal to justify their interference with her decision to have an abortion. But any such intrusion would do very little, if anything, to improve her character or to make her more appreciative of potential human welfare. Hence, it could not be a sufficient reason to interfere with her decision to have an abortion.

Moreover, any decision to have an abortion, even when it is to avoid personal hardship, seems to reveal disrespect for a human being. Now an important moral virtue is respect for the wishes and choices of each and every human being. This includes respect for their hypothetical wishes and choices, for decisions they would make were they able to do so. For example, when the mangled victim of an accident is brought to the emergency room of a hospital, physicians ought to operate only if she consents to this treatment. And if she is unconscious, then what determines the morality of treatment is her presumed consent, what she would have chosen were she capable of making a rational decision. Respect for the wishes and choices of others, including their hypothetical desires and decisions, is normally a requirement for morally right action and is also a morally good character trait. Now to abort a healthy fetus is to disregard the hypothetical choice of the child who would otherwise have been born and would be glad that she was not aborted. Again, to mutilate a corpse expresses disrespect for the deceased because, even if he never thought of this possibility, he would not have chosen to have his body mistreated in this way. To be sure, it is only in an extended sense that one can speak of the hypothetical wishes and choices of beings lacking the psychological capacities required to have desires or make decisions. But it may well be that human nature is such that one cannot cultivate and sustain a vigorous respect for normal adult human beings without nurturing also a respect for beings who could become or have been fully human. Respect for the wishes and choices of others, including their hypothetical desires and decisions, is a character trait essential for sociability. It may even be that widespread respect for each and every human being is a necessary condition for social stability and personal security. However, this is not a sufficient reason for the intervention of others to prevent a woman from having an abortion because to thus frustrate her own desires and decision will do nothing to sustain her disposition to respect the wishes and choices of others. Nor will her individual action of having an abortion do very much, if anything, to undermine such respect in the general public. Therefore, although there are reasons to believe that having an abortion is sometimes morally wrong, none of these reasons is under normal circumstances sufficient to justify the interference of others with the decision of a pregnant woman to have an abortion. Even when her action might be morally wrong, her decision remains private and normatively protected by her human right to privacy.

8.5. The Ideal Scope

Ideally, the law ought to respect the human rights of those subject to it. Presumably, therefore, the scope of the constitutional right to procreational autonomy ought to be defined broadly enough so that it does not permit the state to violate the fundamental moral right to privacy of any individual deciding whether to procreate or prevent procreation. Because procreational decisions are in general private, both the legal right of married individuals to procreate or not procreate and their legal right to prevent or not prevent procreation ought to be defined in very general terms. Thus, the scope of the constitutional right to procreational autonomy ought to be broad enough to permit and protect the liberty of married individuals to make and act on any and all procreational decisions under normal circumstances.

Even granted the existence of the human to privacy, why should the law take this moral right so seriously that it ought to define the constitutional right to procreational autonomy in conformity with it? It is not that the human right to the privacy of decisions is always very important. Many private decisions, like my decision whether to eat lunch on campus or at home today, are trivial. But when a private decision is also momentous, as a procreational decision typically is, the importance of the decision makes it morally urgent for the law to respect the privacy of that decision under normal circumstances. And some procreational decisions are private even when it would be morally wrong to act on them. It follows that the boundaries of the legal right to procreate or not procreate and the legal right to prevent or not prevent procreation ought to be defined more broadly than the corresponding moral procreational rights. Moreover, sometimes the law can give reliable protection for a right only by providing some area of protection beyond the boundaries of that right. For example, the common law as traditionally defined protected the individual from bodily injury or unwelcome bodily contact more securely by adding the tort of assault to the tort of battery. Similarly, the constitutional right to freedom of speech protects the individual's exercise of her human right to freedom of expression more reliably by permitting her to express herself even when she may be speaking in a manner outside the boundaries of her moral liberty to express herself.

Nevertheless, it is probably morally permissible or even morally required to limit the scope of the constitutional right to procreational autonomy by recognizing specific exceptions under special circumstances. Although the married individual ought to have the legal liberty to decide whether to procreate or to prevent procreation in general, there are some ways of exercising this liberty that are so immoral that they ought not to be legally permissible. For example, the husband's constitutional right to procreational autonomy ought not to extend to procreating by raping his wife or preventing procreation by a physical assault upon her person to cause a miscarriage. However, I believe that limiting exceptions to the general right to procreational autonomy should be made only on independent grounds. For example, rape and wife-beating are and ought to be legally prohibited whether or not performed with the intention or effect of procreating or preventing procreation. Thus, the general legal rights to procreate or not procreate and to prevent or not prevent procreation should be limited only when some specific manner of exercising these rights could be justifiably prohibited in nonprocreational contexts.

Just as no constitutional right ought to be unlimited in scope, so none ought to be absolute. Even within its defined limits, the law ought to enable the government to override an individual's right when there really is a sufficient moral justification for doing so. What moral reason or reasons would constitute sufficient grounds for infringing on the privacy of the married individual's procreational decisions? The *European Convention for the Protection of Human Rights and Fundamental Freedoms* attempts to specify the permissible grounds for overriding the human right to respect for private and family life.

> There shall be no interference by a public authority with the exercise of this right except as . . . is necessary in a democratic society in the interests of national security, public safety or the economic well-being of the country, for the prevention of disorder or crime, for the protection of health or morals, or for the protection of the rights and freedoms of others. (Article 8.2)

Should this or some similar formulation define the reasons United States law ought to recognize as grounds for overriding the constitutional right to procreational autonomy? I am not confident that this list of justifying reasons is complete, and I suspect that these reasons are described so vaguely that they could be used to infringe this important right unjustifiably. I would prefer for the courts to retain the recognized general principle that a constitutional right may be overridden only when necessary for some compelling state interest. The words "necessary" and "compelling" are strong enough to prevent almost all arbitrary infringements of the married individual's right to procreational autonomy. At the same time, given our inability to predict what legal issues will arise in the future and under what circumstances, the authority of the courts to identify those state interests that at any given time are compelling and to decide whether in the case before the court they make overriding the individual's constitutional right necessary provides a very useful flexibility.

Some jurists will question my distinction between limiting the constitutional right to procreational autonomy and overriding it. They will justifiably assume that as a moral philosopher I lack an adequate appreciation of how the law really functions. It may be true that a court sometimes reasons that an applicable constitutional right should be overridden in a given case. But this judicial decision, especially if it is a decision upon appeal, will in the future serve as a legal precedent that seems to define an exception to the general right and thereby limit it. Hence, they will argue, for a court to override a right is really to limit that right. I agree that this is often, in fact usually, how the United States legal system functions. But subsequent decisions can and sometimes will sustain the distinction between limiting and overriding the applicable right. It all depends upon how closely the courts follow the judicial reasoning in the original case. Although they often take that decision to limit the right so that it is now irrelevant to similar cases, they sometimes continue to reason that the constitutional right remains applicable but is overridden in later cases.[69] Therefore, I propose that the constitutional right to procreational autonomy should be defined very broadly in general terms, that its core liberties should be limited only when it would be justifiable to prohibit how they are exercised on independent grounds, but that it should be overridden when but only when necessary for some compelling state interest.

---

69 Wellman, *Real Rights*, pp. 259–260.

CHAPTER 9

POSSESSORS OF THE RIGHT TO PROCREATIONAL AUTONOMY

Jurists and philosophers of law agree that there is some sort of a constitutional right to procreational autonomy under United States law, but they disagree about what its precise content is or ought to be. They disagree even more about which classes of persons actually do or ought to possess it. Although my purpose is to discuss all of these legal and moral issues, I have sought to simplify my treatment of them by discussing the scope of the constitutional right to procreational autonomy first before going on to confront the more controversial issues concerning its possessors.

In the previous essay, I have proposed and defended my definition of its scope under existing law and explained what I believe that its scope ought to become in the future. I have argued that the full right to procreational autonomy is a rights-package consisting of two distinct but related rights, the liberty-right to procreate or refrain from procreating and the liberty-right to prevent or not prevent procreation. In addition to its defining bilateral liberty, each of these includes a number of associated legal positions, most notably a claim against interference with the right-holder's exercise of her constitutionally protected liberty. In order to bracket questions about who does or ought to possess this right, I have limited my discussion to the right possessed by married persons only.

Now, I intend to inquire as to what other classes of persons actually possess or ought to possess a constitutional right to procreational autonomy. For example, do unmarried persons, especially homosexual males or single women on welfare, young children or adolescents, intellectually limited or mentally ill persons, and convicted felons also possess a constitutional right to procreational autonomy? And if so, do they possess the full right or is their right more limited in scope? In the end, I will also consider what classes of persons ought to possess this right and whether there are moral justifications for limiting their right more narrowly than that of normal adult married persons.

9.1. Ascriptions of the Right

Who does in fact possess a constitutional right to procreational autonomy? The best way to begin answering this question is by an examination of the opinions of our courts, especially the United States Supreme Court, to see to what classes of persons they have ascribed some such right. The original leading case concerning the constitutional right to procreate was *Meyer v. Nebraska*.[1] Although it did not explicitly mention any constitutional

[1] *Meyer v. State of Nebraska*, 262 U.S. 390 (1923).

right to procreate, it did assert that the liberty protected by the Fourteenth Amendment includes the right "to marry, establish a home and bring up children."[2] This passage clearly links the right to marry with the rights to establish a home and to bring up children. Presumably, these rights include the right to procreate the children the married couple will bring up in their home. Thus, *Meyer* ascribes the right to procreate to married persons.

The first case recognizing a constitutional right to prevent procreation was *Griswold v. Connecticut*.[3] In it the Supreme Court declared unconstitutional a statute forbidding the use of contraceptives even by married persons.[4] That its ascription of the right to use contraceptives, and thereby to prevent procreation, was intended to apply only to married persons is suggested both by the fact that the words "married persons" were italicized in its opinion and that in its reasoning it asked the rhetorical question "Would we allow the police to search the sacred precincts of marital bedrooms for telltale signs of the use of contraceptives?"[5]

In *Eisenstadt v. Baird*,[6] however, the Supreme Court extended the reasoning of *Griswold*, that had first recognized a constitutional right to privacy, to unmarried persons also. The crucial sentence reads, "If the right to privacy means anything, it is the right of the *individual*, married or single, to be free from unwarranted governmental intrusion into matters so fundamentally affecting a person as the decision whether to bear or beget a child."[7] In addition to ascribing the constitutional right at issue to single individuals as well as married persons, this frequently cited sentence is relevant to our purposes in two other ways. First, it generalized the content of the right from merely the right to use contraceptives to a right concerning whether or not to bear or beget a child, that is, a right concerning both procreating and preventing procreation. Second, by emphasizing that this right concerns the decision whether to bear or beget a child, it very clearly recognized a constitutional right to procreational autonomy, a decision free from governmental intrusion.

Does it matter whether the individual, married or single, who is deciding whether to bear or beget a child is an adult? In *Carey v. Population Services International*, the Supreme Court asserted that "... the right to privacy in connection with decisions affecting procreation extends to minors as well as to adults."[8] It based this conclusion upon its previous decision in *Planned Parenthood of Central Missouri v. Danforth*.[9] Although its decision clearly ascribed the constitutional right to procreational autonomy to minors, *Carey* was not intended to extend this right to very young children. It noted that "*Planned Parenthood*, however, "does not suggest that every minor, regardless of age or maturity, may give effective consent for termination of her pregnancy."[10] Hence, it leaves open the question as to whether its decision applies to minors who are not yet adolescents.

The Supreme Court of California in *Conservatorship of Valerie N.* was asked to determine whether a statute precluding the sterilization of a severely retarded conservatee

---

[2] *Ibid.* at 399.
[3] *Griswold v. Connecticut*, 381 U.S. 479 (1965).
[4] *Ibid.* at 480, 485.
[5] *Ibid.* at 480, 485–486.
[6] *Eisenstadt v. Baird*, 405 U.S. 438 (1972).
[7] *Ibid.* at 453, italics in original.
[8] *Carey v. Population Services International*, 431 U.S. 678 (1977) at 693.
[9] *Planned Parenthood of Central Missouri v. Danforth*, 428 U.S. 52 (1976).
[10] *Carey*, footnote 16 at 693.

in all circumstances was valid.[11] It described the constitutional right at issue as "the right of procreative choice" and as "the right of women to exercise procreative choice 'as they see fit.'"[12] The opinion of the Court then ascribed this right to incompetent developmentally disabled women.[13] How, if at all, its judicial reasoning might apply to severely retarded men was not considered.

Finally, the constitutional right to procreational autonomy was given its most explicit definition in *Davis v. Davis*.[14] Although this case is a binding precedent only in Tennessee, it is grounded squarely upon an analysis of prior decisions of the United States Supreme Court. The case began as a divorce action in which Junior Davis and Mary Sue Davis disagreed about who should have "custody" of the seven "frozen embryos" stored in a Knoxville fertility clinic.[15] The Supreme Court of Tennessee observed that:

> The United States Supreme Court has never addressed the issue of procreation in the context of *in vitro* fertilization. Moreover, the extent to which procreational autonomy is protected by the United States Constitution is no longer entirely clear.... For the purposes of this litigation it is sufficient to note that, whatever its ultimate constitutional boundaries, the right of procreational autonomy is composed of two rights of equal significance—the right to procreate and the right to avoid procreation.[16]

It then ascribed this right to each of the two gamete-providers whose sperm and ova, respectively, were used to produce the pre-embryos that could be used for *in vitro* fertilization. Although the parties in this case were previously married, the reasoning of the Court would seem to apply generally to all gamete-providers.

Given these precedents, who possesses the constitutional right to procreative autonomy? The line of cases beginning with *Meyer* and *Griswold*, merging in *Eisenstadt* and then continuing through *Carey* and *Davis* clearly establish that adult married persons, at least under normal circumstances, possess this right. In addition, *Eisenstadt* shows that many unmarried adults also possess the right to procreational autonomy. *Carey* extends this constitutional right to adolescents, but leaves open the age at which minors acquire it. Similarly, *Conservatorship* clearly shows that developmentally disabled or severely retarded women possess at least the constitutional right to prevent procreation, but leaves unclear whether they also possess the right to procreate.

9.2. Reasoning of the Courts

These cases suggest several important questions concerning the possession of the constitutional right to procreational autonomy. Are there some classes of single individuals, for example homosexual males or women on welfare, who do not possess this right? Do

---

[11] *Conservatorship of Valerie N.*, 707 P. 2nd 760 (Cal. 1985), at 762.
[12] *Ibid.* at 771, 772.
[13] *Ibid.* at 773–774.
[14] *Davis v. Davis*, 842 S.W. 2nd 588 (Tenn. 1992).
[15] *Ibid.* at 589.
[16] *Ibid.* at 601.

men and women equally possess the right to prevent procreation? Do mentally retarded women possess the right to procreate as well as the right to prevent procreation? At what age do minors acquire the constitutional right to procreational autonomy? Could someone forfeit this right by one's criminal actions? Let us try to answer these questions by an examination of the reasoning in which the courts have appealed to the constitutional grounds of the right to procreational autonomy.

In *Bellotti v. Baird*,[17] often referred to as *Bellotti II*, the United States Supreme Court derived the right to procreational autonomy from the constitutional right to privacy. It cited *Roe v. Wade*, *Doe v. Bolton*, and *Planned Parenthood of Central Missouri v. Danforth*, all cases that hinged on the right to privacy, as evidence of a woman's constitutional right to choose to terminate her pregnancy.[18] At issue was the validity of a Massachusetts statute that did not permit pregnant minors, mature or immature, to petition a court for permission to have an abortion without any parental consultation. The Supreme Court reasoned that this requirement would impose an undue burden upon the exercise by minors of the right to seek an abortion. It concluded that "under state regulation such as that undertaken by Massachusetts, *every minor* most have the opportunity—if she so chooses—to go directly to a court without first consulting or notifying her parents."[19] Presumably, the reasoning of the court in this case implies that all minors, whatever their age or degree of maturity, possess the constitutional right to procreational autonomy.

It might seem, however, that this presumption is inconsistent with the subsequent reasoning of the Supreme Court in *H. L. v. Matheson*.[20] In this case, the Court upheld the constitutionality of a Utah statute requiring a physician to notify the parents of a minor before performing an abortion. It considered only the facial constitutionality of this statute as applied to a minor daughter who is living with and dependent upon her parents, is not emancipated and has made no claim to her maturity.[21] I would argue, however, that the reasoning of the Supreme Court did not assume that this limited class of minors has no right to procreational autonomy, but that their right is not absolute. Hence, the Court concluded that any burden this statute might impose upon its exercise was justified by the important state interests in encouraging a minor to seek the help and advice of parents in making a very important decision, in furthering parental authority and family integrity, providing an opportunity for the parents to supply essential information to the physician, and protecting potential life.[22] Therefore, taken together, these cases do show that every minor possesses a constitutional right to procreational autonomy.

In *Davis v. Davis*,[23] the Supreme Court of Tennessee also derived the right to procreational autonomy from the right to privacy. "Here, the specific individual freedom in dispute is the right to procreate. In terms of the Tennessee state constitution, we hold that the right of procreation is a vital part of an individual's right to privacy. Federal law is to the same effect."[24] More specifically, the dispute was between two recently

---

[17] *Bellotti v. Baird*, 443 U.S. 622 (1979).
[18] *Ibid.* at 638–639.
[19] *Ibid.* at 647, emphasis added.
[20] *H. L. v. Matheson*, 450 U.S. 398 (1981).
[21] *Ibid.* at 407.
[22] *Ibid.* at 411–413.
[23] *Davis v. Davis*, 842 S.W. 2nd 588 (Tenn. 1992).
[24] *Ibid.* at 600.

divorced individuals concerning the disposition of their frozen embryos. Mary Sue Davis wanted to use them to procreate, but Junior Lewis Davis desired to prevent her from procreating.

The Supreme Court of Tennessee reasoned that

> For the purposes of this litigation it is sufficient to note that, whatever its ultimate constitutional boundaries, the right of procreational autonomy is composed of two rights of equal significance—the right to procreate and the right to avoid procreation. Undoubtedly, both are subject to protections and limitations.[25]

Accordingly, the Court balanced the interests of the individual parties to this case and concluded that given their particular circumstances, the right to prevent procreation of Junior Davis outweighed the right to procreate of Mary Davis. It noted, however, that under different circumstances, the balance of interests would usually imply a different holding.[26] Of special relevance to our subject is the fact that it was Junior Davis, a male, to whom the Court ascribed the right to prevent procreation. It distinguished this case from the line of abortion cases that had held that the decision whether to terminate a pregnancy belongs exclusively to the pregnant woman.

> The equivalence of and inherent tension between these two interests are nowhere more evident than in the context of *in vitro* fertilization. None of the concerns about a woman's bodily integrity that have previously precluded men from controlling abortion decisions is applicable here .... As they stand on the brink of potential parenthood, Mary Sue Davis and Junior Lewis Davis must be seen as entirely equivalent gamete-providers.[27]

Because the reasoning of the Court assumes that men and women equally possess the constitutionally protected interests in procreating and not procreating, it implies that both possess the constitutional rights to procreate and to prevent procreation.

Although the courts have usually derived the right to procreational autonomy from the constitutional right to privacy, there are other cases that hinge upon the right to the equal protection of the laws. Most notably, this is true of the leading case of *Skinner v. Oklahoma*.[28] At issue was the constitutionality of Oklahoma's Habitual Criminal Sterilization Act that authorized the involuntary sterilization of individuals convicted of three felonies. The Opinion of the Supreme Court began by asserting, "This case touches a sensitive and important area of human rights ... the right to have offspring."[29] It then reasoned that the Oklahoma statute violated the Equal Protection Clause because

> When the law lays an unequal hand on those who have committed intrinsically the same quality of offense and sterilizes one and not the other, it has made as invidious a discrimination as if it had selected a particular race or nationality for oppressive teatment .... Sterilization of those who have thrice committed

---

[25] *Ibid.* at 601.
[26] *Ibid.* at 603–604.
[27] *Ibid.* at 601.
[28] *Skinner v. Oklahoma*, 316 U.S. 535 (1942).
[29] *Ibid.* at 536.

> grand larceny, with immunity for those who are embezzlers, is a clear, pointed, unmistakable discrimination.[30]

Thus, although *Skinner* assumes rather than proves the existence of a fundamental right to have offspring, it shows that habitual criminals continue to possess the right to procreate because one does not forfeit this right even by one's repeated criminal actions.

In *North Carolina Association for Retarded Children v. State of North Carolina*,[31] a United States District Court upheld, for the most part, the constitutionality of a North Carolina statute that authorized both voluntary and involuntary sterilizations. The reasoning of the Court did not assume that persons who are mentally ill or mentally retarded lack the right to procreational autonomy; rather it presupposed that they do possess this right because it subjected the disputed statute to strict scrutiny.

> Moreover, the classification is itself narrowed as to impact so that, as we interpret it, only mentally retarded persons who are sexually active, and unwilling or incapable of controlling procreation by other contraceptive means, *and* who are found to be likely to procreate a defective child, *or* who would be unable because of the degree of retardation to be able to care for a child, may be sterilized. The legislative dual purpose—to prevent the birth of a defective child or the birth of a nondefective child that cannot be cared for by its parent—reflects a compelling state interest and the classification rests upon a difference having a fair and substantial relation to the object of the legislation and does not, therefore, violate the Equal Protection Clause of the Fourteenth Amendment of the Constitution of the United States.[32]

This presumption becomes explicit in another passage.

> We agree with the United States that the right to procreate is a fundamental right.... To sustain this statute against substantive due process challenge it must be found that the state's interest is "compelling." We interpret Article 7 as narrowly drawn to express only the legitimate State interest of preventing the birth of a defective child or the birth of a nondefective child that cannot be cared for by its parent, and that so viewed, the State's interest rises to the dignity of a compelling one.[33]

Clearly, then, this line of reasoning implies that mentally retarded persons possess the right to procreational autonomy.

Finally, in *Maher v. Roe*,[34] the Supreme Court upheld the constitutionality of a regulation of the Connecticut Welfare Department that limited state Medicaid benefits for first trimester abortions to those that are "medically necessary." Two indigent women complained, among other things, that this regulation violated the Equal Protection Clause because the Connecticut Welfare Department funded the medical expenses for childbirth but not for abortion. Once more, the reasoning of the Court did not deny that women on

[30] *Ibid.* at 541.
[31] *North Carolina Association for Retarded Children v. State of North Carolina*, 420 F. Supp. 451 (1976).
[32] *Ibid.* at 457–458, italics in original.
[33] *Ibid.* at 458.
[34] *Maher v. Roe*, 432 U.S. 464 (1976).

welfare possess the right to procreational autonomy. Instead, it distinguished this regulation from those invalidated in its previous abortion decisions.

> The Connecticut regulation places no obstacles absolute or otherwise—in the pregnant woman's path to an abortion. An indigent woman who desires an abortion suffers no disadvantage as a consequence of Connecticut's decision to fund childbirth; she continues as before to be dependent on private sources for the service she desires. The State may have made childbirth a more attractive alternative, thereby influencing the woman's decision, but it has imposed no restriction on access to abortions that was not already there. The indigency that may make it difficult—and in some cases, perhaps, impossible—for some women to have abortions is neither created nor in any way affected by the Connecticut regulation. We conclude that the Connecticut regulation does not impinge upon the fundamental right recognized in *Roe*.[35]

Here, the reasoning of the Supreme Court implicitly reaffirmed the possession by indigent persons on welfare of the right to procreational autonomy.

Thus, an examination of the reasoning of the courts partially, but not entirely, answers the questions suggested by those cases in which the courts have ascribed the right to procreational autonomy to specified classes of individuals. Men and women equally possess this general constitutional right, although their possession of the more specific rights this implies, such as the right to choose an abortion, may differ. Women on welfare, mentally retarded persons and minors whether mature or immature possess the right to procreational autonomy, but their exercise of this right may be limited in some circumstances. And habitual criminals have not lost the right to procreational autonomy by their repeated violations of the law.

9.3. Potential Judicial Decisions

Although the courts have decided many of the legal issues concerning the possessors of the constitutional right to procreative autonomy, they have left others undecided. Moreover, some of their decisions are controversial enough so that they might be reconsidered in the next few years. Hence, there is the possibility, in some cases the probability, that various classes of claimants will challenge the constitutionality of new attempts by the several states to deny them this right. Although any attempt to predict how the courts will decide such cases might be so speculative as to be virtually useless, it will be illuminating to discuss how important moral reasons bear on several potential judicial decisions. This requires us to go beyond a consideration of who actually does possess the constitutional right to procreative autonomy and to decide who ought on moral grounds to possess this legal right.

Adult married persons are paradigm examples of those who do and ought to possess this right. That they, if anyone, do possess this right is amply demonstrated by the close connection United States courts found between marriage and the right to procreate in the

[35] *Ibid.* at 474.

leading case of *Meyer* and the right to prevent procreation in *Griswold.* Even though *Eisenstadt v. Baird* and subsequent judicial decisions later extended these rights to unmarried individuals, I cannot think of any court that has considered denying these rights to normal adult married persons. I have argued elsewhere[36] that married adults ought to possess this legal right, at least under normal circumstances, in order to protect their human right to procreational autonomy and to respect their human right to privacy. This leaves open, however, whether other classes of individuals ought to possess the full constitutional right to procreational autonomy.

(1) Ought unmarried adults in stable heterosexual relationships to possess the constitutional right to procreational autonomy. I believe that they ought to possess this legal right because this would recognize and help to protect their analogous human right. The human right to procreational autonomy is a rights-package consisting of a basic moral right to choose whether to procreate or refrain from procreating and a basic moral right to choose whether to prevent or not prevent procreation. The moral reasons that ground these two constituent rights apply to unmarried adults in stable heterosexual relationships just as they do to married adults.

Consider first the human right to prevent procreation. Its defining core consists of the bilateral moral liberty to prevent procreation or not prevent procreation as one chooses. The moral liberty to prevent procreation by preventing the procreative process from beginning, as by the use of contraceptives, is probably an innocent liberty simply because a liberty is the absence of a contrary duty and I can think of no moral reason sufficient to impose a general duty to procreate. However, the liberty to terminate this process seems to be suspect because aborting a fetus often destroys the potentiality for a healthy and happy human life. Nevertheless, it is morally justified when it is necessary to prevent the birth of a seriously defective child or when the birth of a child would impose severe hardship upon one or both parents or when the parents are unwilling or unable to care for the child adequately. The moral liberty to refrain from preventing procreation is also in general an innocent liberty because there is no moral reason sufficient to impose a duty not to procreate under normal circumstances. Now whether an adult is married or unmarried makes no difference to the relevance of any of these moral considerations.

Nor does one's marital status seem relevant to the moral reasons that ground the human right to procreate or refrain from procreating as one chooses. Because there is no moral duty to procreate, the moral liberty to refrain from procreating is an innocent liberty. And the moral liberty to procreate is in general an innocent liberty also because under normal circumstances there is no moral reason sufficient to impose a moral duty to refrain from procreating. There are, however, exceptions to this general moral liberty, at least the three exceptions mentioned above. Therefore, I believe that these moral reasons are sufficient to ground a general moral right to procreational autonomy of adults who are either married or living in stable heterosexual relationships.

There might, however, be additional moral reasons that would imply additional exceptions to this general human right. I very much doubt that there is any moral duty to refrain from preventing procreation merely because one is unmarried. If anything, one who takes marriage to be an important moral value would wish to reduce the birth of

---

[36] In the previous essay.

illegitimate children. More plausible is the view that being unmarried somehow reduces or eliminates the moral liberty to procreate possessed by those who are married. Why might this be so?

Article 23 of the *International Covenant on Civil and Political Rights* reads in part:

1. The family is the natural and fundamental group unit of society and is entitled to protection by society and the State.
2. The right of men and women of marriageable age to marry and to found a family shall be recognized.

By asserting "the right," in the singular rather than the plural, "to marry and to found a family," this passage suggests that the human right to found a family, presumably by procreation, is conditional upon marriage and that this dependence is somehow based upon the fact that the family is the fundamental group unit of society. Some such notions probably lie behind many decisions of our courts that appeal to the social importance of marriage and family values.[37]

Why is the family entitled to protection by society and the State? This social institution deserves to be maintained and supported by our legal system because it is of great value in several ways to both its family members and to society as a whole. It is a basic economic unit for the production and distribution of goods and services. This was most clearly true in the preindustrial era when most people lived on family farms and even during the transitional period of cottage industries. But it remains true even today in our highly industrialized societies. Income earned outside the home is spent to meet the needs of the family, and the goods and services purchased are distributed to the members of the family as the parents decide. Within the family there is an informal division of labor that promotes the efficient production of goods, such as casseroles or homemade bookshelves, and the provision of services, such as changing diapers, making beds, or mowing lawns.

The family is the primary institution for providing social security to its members. When one becomes ill, other members of the family will care for one. Only in cases of serious illness will physicians, nurses, or hospitals be needed; and even then their function is usually to supplement family care rather than to replace it. If someone becomes unemployed, he or she will be sheltered, fed, and clothed by the other family members. State unemployment benefits, if any, merely serve to enable the family to provide the necessary social security. The family provides for the basic needs of those who are too young to care for themselves, and in old age one relies in the first instances upon one's family to sustain one.

The family is a small social grouping that is most conducive to lasting intimate personal relationships. The fact that the members of a family live together and share so much of their lives promotes close personal relationships that give meaning and emotional satisfaction to individual lives. Although these are not always loving and nurturing, they typically are, and they are much more difficult to create and sustain in larger and more impersonal groups.

Finally, and most relevant to our purposes, the family provides for the needs of the growing and maturing child. Parents do more than feed, clothe, shelter, and protect their children from harm. They educate their children by teaching them the basic skills

---

[37] For example, *Meyer v. Nebraska*, 262 U.S. 390 (1923); *Skinner v. Oklahoma*, 316 U.S. 535 (1942); and *Prince v. Massachusetts*, 321 U.S. 158 (1944).

they will need to live satisfying lives and develop their moral character to enable them to become morally responsible adults and useful citizens. When necessary, they arrange for professional medical care, public or private schooling and integration into the larger society for them. In all of these ways the family is a fundamental social institution of immense value to its members and to society.

Now a family consisting of unmarried adults living in stable heterosexual relationships can and typically does fulfill all of these valuable social functions. It is a basic unit for the production and distribution of goods and services. It provides social security to its members. It is conducive to the creation and enjoyment of loving intimate personal relationships. And because this sort of family includes both female and male caregivers and provides both male and female role models, it seems fully capable of nurturing and raising their children. Why might it matter that the parents are unmarried?

Well, it might matter because married parents might have a more stable relationship than unmarried partners. The fact that they have been willing to commit themselves by a promise of lifelong fidelity indicates a stronger and deeper commitment to one another than is usual by those unwilling enter into a legally binding marriage. Although divorce is possible and not uncommon, it is legally possible only within the limits prescribed by the State. This stability might be essential for the family to ensure social security to its members, especially its children. Beyond that, knowing that one's parents are married may well be important to the emotional security required for a child to grow up with full self-confidence and self-respect.

Whether married persons do in fact enjoy a significantly more stable relationship than unmarried partners is an empirical question upon which I am not qualified to voice any expert opinion. However, the prevalence of divorce in our society today and the fact that many undivorced couples separate for long periods or even permanently casts some doubt upon this assumption. In any event, procreation is legally permissible at least from the very beginning of a marriage, and any general legal presumption of marital stability is highly questionable. The denial of the basic human right to procreational autonomy would require a much firmer foundation than this.

A much more plausible reason for conditioning the right to procreate upon marriage is that marriage is a legally defined relationship. Hence, it is much easier to determine whether a man and a woman have met the formal requirements for marriage than whether they have entered into a stable heterosexual relationship. How long must they have lived together or by what signs must they have demonstrated their long-term commitment to one another before the law should recognize their union as stable? More importantly, although marriage is a contract insofar as the consent of both parties is required for it to be valid, the rights and duties of the family members are defined by the law and cannot be terminated or redefined by the partners.[38] The rights and duties of unmarried partners, on the other hand, remain controversial and are less clearly defined by the family law of our several states. Therefore, it is much easier to fix the responsibilities of married parents concerning the care of their children and the rights of the children holding against their parents and against the State as *parens patriae*.[39] Hence, marriage serves to provide a much more secure protection to any children who may result from the exercise of the

[38] See *Maynard v. Hill*, 125 U.S. 190 (1887) at 211.

[39] Compare *Hewitt v. Hewitt*, 394 N.E. 2nd 1204 (1979) at 1207–1208.

constitutional right to procreation. One might argue that the morally compelling State interest in protecting children from harm would justify limiting the possession of the right to procreate to married persons.

Granted that the state interest in protecting children from serious harm is morally compelling, limiting the constitutional right of procreational autonomy to married adults would do little to promote it. For one thing, adults are permitted to marry without any test or evaluation of their willingness or ability to carry out their parental responsibilities, and child neglect or abuse by married parents is far from rare. For another, the prohibition of procreation outside of marriage has not been and in the future would not be an effective means of preventing the birth of illegitimate children. A much better way to protect children would be to impose the same legal duties of childcare upon unmarried parents as those that are now clearly defined in the law of marriage.

I have argued that unmarried adults in stable heterosexual relationships have the same human right to procreational autonomy as do married adults. Therefore, only some morally compelling state interest could justify denying them the analogous legal right. Because the two most plausible justifications of this sort fail, I conclude that unmarried adults in stable heterosexual relationships ought to possess the constitutional right to procreational autonomy. The same conclusion might well be grounded upon their basic moral rights to privacy and to equitable treatment, but I will postpone discussion of these rights for the moment.

(2) Ought adults in stable same-sex relationships to possess the constitutional right to procreational autonomy? I shall argue that their human right to privacy implies that they should. The human right to privacy is a rights-package consisting of the four human rights to the privacy of one's spaces, information about one, one's experiences, and one's decisions. It is the last of these that is relevant here. The defining core of this human right to the privacy of one's decisions is a moral claim against others that they not intrude into any of one's private decisions by preventing or seriously hindering one from acting as one chooses. A private decision is one that affects the life of the agent for better or worse and such that no one else has a sufficient reason to interfere coercively.

The primary moral ground of this fundamental moral right in general is the fact that interference with one's decision is experienced as an alien and unwelcome intrusion into one's life and threatens to disrupt one or more of the projects that give meaning to that person's life. This claim against intrusion is stronger than usual when one is acting on a personal decision, one such that how one ought to act depends in large measure upon one's feelings, goals, value judgments or projects, and factors that vary greatly from person to person. And this claim against interference is strongest when one is acting on a momentous decision, one that greatly affects one's life as a whole for better or worse. Decisions concerning procreation are typically both personal and momentous because whether one procreates or prevents procreation makes a very great difference to the things one values most and to many of the ongoing projects that give meaning to one's life. These moral reasons apply to adults in stable same-sex relationships just as much as they apply to adults in stable heterosexual relationships. Therefore, adults living in stable same-sex relationships possess a human right to privacy that imposes a *prima facie* moral claim upon the state not to interfere with any of their procreational choices by denying them the constitutional right to procreational autonomy.

Unless, of course, the State, has a sufficient reason to prevent them from acting on their procreational decisions. One reason might be that a same-sex family does not adequately serve essential family values.[40] One must not, of course, define family values rigidly in terms of the traditional heterosexual family, for that would be to beg the question. Rather, let us look to the essential functions of the family as a social institution. There is no doubt that a same-sex family can and typically would produce and distribute goods and services for its members, provide social security to them, promote lasting intimate personal relationships, and provide for the needs of any children. The question remains whether it would perform these functions adequately? Well, why not? In contrast with a heterosexual family, in most cases only one of the partners would be biologically related to any of "their" children. This might well reduce the love and commitment of the unrelated partner to the child. Thus, the family might have only one fully committed caregiver and be less effective in creating and maintaining lasting intimate relationships. And given the reliance our legal system places upon biological relationships to identify the persons responsible for the care and support of a child, this would undermine the value of a same-sex family as an institution that provides social security to its children.

Whether same-sex families would be significantly less effective, if at all, in promoting essential family values in these ways is an empirical question. This sort of nontraditional family has probably not existed in sufficient numbers for long enough to enable sociologists and social psychologists to establish any highly reliable conclusions in this regard. Therefore, the allegation that same-sex families do not adequately promote family values is at best speculative and does not provide a firm enough foundation to be morally sufficient for limiting the human right to privacy of adults in stable same-sex relationships.

Another reason that might justify denying the constitutional right to procreational autonomy to adults in stable same-sex relationships might be that they would be unfit parents. Obviously, there could not be both a female and a male parent in any same-sex family. In our society, a man and a woman normally provide childcare in different amounts and of differing kinds. But one of these complementary kinds of care, both of which are presumably needed by the child, will arguably be lacking for the child of a same-sex couple. Also, the child in a same-sex family would not have both male and female role models. Hence, a male child would lack an appropriate role model in a lesbian family and a female child would lack one in a homosexual family. Even if this line of reasoning fails to distinguish between sex and gender, as some feminists argue, these are very highly correlated in our society and people are generally expected to conform to the traditional norms of sex and gender. Fortunately, empirical research does not confirm any of these defects in same-sex families.[41]

In our legal system, parental competence is, and morally ought to be, measured by the best interests of the child. The State of Florida justified its statute denying the right to adopt to lesbian and homosexuals on the ground married heterosexual couples best serve the best interests of any child.[42] Whether rigidly maintaining the traditional female and male role models is advantageous to, much less essential for, the best interests of children

---

[40] But see M. Strasser, "Sex, Law, and the Sacred Precincts of the Marital Bedroom: On State and Federal Right to Privacy Jurisprudence", *Notre Dame Journal of Law, Ethics and Public Policy* 14 (1998), 771–772, 776–777, 789.

[41] B. Fitzgerald, "Children of Lesbian and Gay Parents: A Review of the Literature", *Marriage and Family Review* 29 (1999), 61–68.

[42] *Lofton v. Kearney*, 157 F. Supp. 2nd 1372 (S.D.Fla. 2001) at 1383.

is controversial. Whether this is even possible in our society where large numbers of women are now breadwinners and where men are increasingly sharing the responsibilities of childcare is dubious. As a New Mexico Court of Appeals observed in a dispute between two women concerning custody and visitation rights:

> In short, the issue before the court is not the nature of the parent's sexual activities, if any, but whether and how these activities affect the child, if in fact they do. This is a factual issue that must be considered and resolved on specific evidence concerning the effect, if any, of the activity upon the children; it cannot be resolved as a matter of law based on the perceived morality or immorality of the parent's conduct.[43]

Because, I share the skepticism of the court that same-sex parents are always or even usually unfit parents, I believe that they ought not to be denied the constitutional right to procreational autonomy as a matter of law. There is little or no empirical evidence to justify any such general limit upon the possession of this fundamental legal and moral right. Whether some individual lesbian or homosexual adult would be an unfit parent ought to be determined, as it is in the case of heterosexual persons, by factual information about that individual.

A third reason that might justify denying the constitutional right to procreational autonomy to adults living in stable same-sex relationships is that this is necessary to preserve our public morality.[44] Justice Powell, concurring in *Zablocki v. Redhail*, asserted "The State, representing the collective expression of moral aspirations, has an undeniable interest in ensuring that its rules of domestic relations reflect the widely held values of its people."[45] Although he does not explain why this is so, his assertion is very plausible. For one thing, a coherent and widely accepted public morality it probably essential for that social solidarity required for a peaceful society in which the legal system is obeyed out of respect for the law rather than the mere threat of force. Also, the legal system can most effectively perform its function of resolving conflicts between the parties before the court by appealing to values they may be presumed to share. One can argue, therefore, that our legal system ought to deny the constitutional right to procreational autonomy to adults living in same-sex relationships in order to preserve our moral ideal of heterosexual marriage and to conform to the public condemnation of homosexual and lesbian sexual relationships.

Granted the importance for social stability and respect for the law of the widely held moral values of our people, one wonders whether granting the possession of the constitutional right to procreational autonomy to adults in stable same-sex relationships would seriously damage our public morality. Justice Blackmun, dissenting in *Bowers v. Hardwick*, argued

> Reasonable people may differ about whether particular sexual acts are moral or immoral, but "we have ample evidence for believing that people will not abandon morality, will not think any better of murder, cruelty and dishonesty, merely

[43] A.C.*v. C.B.*, 829 P. 2nd 660 (N.M.App. 1992) at 664–665.
[44] *Lofton* at 1382.
[45] *Zablocki v. Redhail*, 434 U.S. 374 (1978) at 399.

> because some private sexual practice which they abominate is not punished by the law."[46]

To be sure, it would be harmful for the law to ignore our most valuable public norms of sexual conduct. It ought to support the moral duty to procreate responsibly, respecting the moral rights of one's sexual partner, and with concern for the needs of any child that will be born. This much is an essential interest of our society and its individual members. But I doubt that the condemnation of same-sex relationships is equally vital to the moral health of our society. As Justice Blackmun also argued

> The fact that individuals define themselves in a significant way through their intimate relationships with others suggests, in a Nation as diverse as ours, that there may be many "right" ways of conducting those relationships, and that much of the richness of a relationship will come from the freedom an individual has to *choose* the form and nature of these intensely personal bonds.[47]

To my mind, this is a strong argument for the conclusion that our legal system ought not to attempt to impose the sexual mores of the majority upon the significant minority of those for whom choosing a different sexual orientation is a momentous private decision. This violation of their human right to privacy is not necessary in order to protect what is essential in our public morality.

An additional reason that might justify denying the constitutional right of procreational autonomy to adults living in same-sex relationships is to prevent the birth and nurture of seriously disadvantaged persons. After all, the State has a moral duty as parens patriae to intervene in a family to protect the best long-term interests of a child, born or unborn. Science has not yet shown whether homosexual or lesbian sexual orientation is determined by genetic inheritance or environmental factors. Either way, any child procreated by an adult homosexual or lesbian and raised in a same-sex family might have a higher probability than the normal child of becoming a homosexual or lesbian adult. And it can hardly be denied that such persons are being seriously disadvantaged in our society. It is harder for them to find and maintain their sexual identity. It is harder for them to find compatible sexual partners and to sustain intimate personal relationships. And they will confront the moral disapproval and social discrimination of many members of our society.

I cannot deny that lesbians and homosexuals are at a disadvantage in our society and suffer psychological and economic harm as a result. I doubt, however, that these harms are so very serious as to justify the denial of the human right to procreate to same-sex couples. In fact, adolescents and adults with same-sex orientations are increasingly able to find compatible sexual partners and to sustain satisfying interpersonal relationships. Were this not so, the question as to whether adults in stable same-sex relationships ought to possess the constitutional right to procreational autonomy would be of little import to our society. Moreover, much of the hardship suffered by lesbians and homosexuals is a result of moral prejudice, not of morally justified disapproval. If at all possible, the law ought to encourage the current trend toward a more enlightened sexual morality, not reinforce cruel and

---

[46] *Bowers v. Hardwick*, 478 U.S. 186 (1986) at 212, quoting H. L. A. Hart.

[47] *Bowers* at 205, italics in original. In *Lawrence v. Texas*, 539 US __ (2003), the United States Supreme Court has recently reversed *Bowers* on similar grounds.

harmful sexual stereotypes. Therefore, preventing the birth of children who may become lesbians or homosexuals is not a reason sufficient to morally justify denying the constitutional right of procreational autonomy to all adults living in stable same-sex relationships.

Nevertheless, it might be easier to justify denying this right to males than to females living in same-sex relationships. While a lesbian can procreate either by having sexual intercourse with a cooperative male or by artificial insemination from donor, a homosexual can procreate only by using the services of a surrogate mother. Although the law probably ought not to prohibit the use of surrogate mothers to heterosexual couples in cases where the gestational mother is a surrogate for the woman who would bear the child were she not infertile, it might be morally permissible for the law to discourage or even prohibit the morally worrisome use of surrogate mothers where the infertility of a heterosexual couple is not at stake. If it seems unfair to deny homosexuals the right to procreate when lesbians are permitted to enjoy it, one must remember that procreational decisions are typically much less momentous for males than for females. This is not merely or even primarily because in our society childbearing and nurturing are less central to their self-images. Unlike a female, a male can never give greater meaning to his life by carrying a fetus to term and delivering a child. Nor can any male ever need the freedom to decide whether or not to have an abortion. Therefore, the State interest in minimizing the use of surrogate mothers might be sufficient to justify denying the constitutional right of procreational autonomy to adult men living in stable homosexual relationships.

I doubt that this is so. Although procreational decisions are typically much less momentous for males than for females, implying that their claim against society not to interfere is usually less strong than for females, this is not always the case. Gender is not invariably tied to sex. Some men desperately want to parent a child and would derive immense satisfaction from caring for their offspring, while many women prefer the unencumbered pursuit of a career to the more traditional social role of mothering. And even when procreational choices may be less momentous for a male, they may still reach the threshold of momentousness. Therefore, a public policy of reducing the use of surrogate mothers is not a moral reason sufficient to justify denying possession of the constitutional right to procreational autonomy to all homosexuals adults living in stable same-sex relationships.

My conclusion is that in general adults living in stable same-sex relationships ought to possess the right to procreational autonomy in the constitutional law of the United States. This is morally required by their human right to privacy as well as their human right to procreational autonomy. Whether some individuals in this class ought to be denied this legal right because of their special circumstances should be decided on a case-by-case basis.

(3) Ought single adults, adults not living in any stable relationship with a partner, to possess the constitutional right to procreational autonomy? Presumably, they should if they possess the analogous human right. Now the grounds of the bilateral liberties that define the core of the basic moral right to procreational autonomy are simply the absences of any general moral duties to refrain from procreating or to procreate. Hence, adults, whether married or unmarried but living in stable relationships with a partner, possess the general moral liberties to procreate or not procreate as they choose and to prevent or not prevent procreation as they choose. Because these grounds would appear to apply to single adults just as they do to adults living in stable relationships, it would seem that single adults possess the same general human right to procreative autonomy that adults living with partners possess.

However, there are exceptions. One who is unable to provide adequate care for a child does have a moral duty not to procreate and by implication not to refrain from preventing procreation. Are single persons generally unable to care adequately for their children? Scientific research suggests that there is no need for a child to have both female and male caregivers or for a family to provide both male and female role models.[48] Still, some might argue that it takes two persons to provide adequate care for a child, especially a very young child. A couple can divide the family roles of breadwinner and child-caregiver. It need not be the case that one person earns the income necessary to provide for the economic needs of the family while the other performs most of the household work. Each may share in both tasks. However, arguable each of these two functions is so demanding that no one person could perform both adequately. And since both are required for rearing a child, very few if any single persons are able to provide adequate childcare. Therefore, in general single adults seem not to possess the full human right to procreational autonomy. They may have the moral liberties to prevent procreation and not to procreate, but they may not be at liberty to choose to procreate or refrain from preventing procreation.

Once more one finds a plausible argument resting upon a questionable premise. Why assume that most single adults are typically incapable of providing for both the economic needs of the family and the various needs of the growing child? To be sure, social scientists have shown that on average children raised in single-parent families do not fare as well as those reared in two-parent families.[49] However, this may be a consequence of the low family income more than the absence of a second parent.[50] Still, the fact that most states permit single adults to adopt children suggests that state legislatures and courts do not always find this a serious disadvantage to the child. Experience shows that many single parents are excellent caregivers and that many others provide care for their children that is more than adequate. Until there is solid evidence to prove that single adults are usually incapable of providing adequate childcare, one should assume that, like other adults, they possess the human right to procreational autonomy. Because the law ought to recognize and protect the human rights of those subject to it, this is one important moral reason to conclude that they ought to possess the analogous constitutional right.

Another reason to support this conclusion is that to deny them the constitutional right to procreational autonomy, while conferring it upon other adults, would be to violate their human right to equitable treatment. The human right to equitable treatment is a fundamental moral claim-right not to be treated less well than others who are similarly situated unless there is a justicizing reason to do so.[51] William Frankena first distinguished justicizing or just-making reasons, reasons that show some action or policy to be just, from other sorts of right-making reasons, reasons that justify without rebutting the claim that it is unjust.[52]

---

[48] D. B. Downey, J. W. Ainsworth-Darnell, and M. J. Dufur, "Sex of Parent and Children's Well-Being in single-Parent Households", *Journal of Marriage and the Family* 60 (1998), 889, 891.

[49] M. J. Carlson and M. E. Corcoran, "Family Structure and Children's Behavioral and Cognitive Outcomes", *Journal of Marriage and Family* 63 (2001), 779.

[50] *Ibid.* p. 780; see also W. H. Jeynes, "The Effects of Several of the Most Common Family Structures on the Academic Achievement of Eight Graders", *Marriage and Family Review* 30 (2000), 91.

[51] See C. Wellman, *Welfare Rights* (Totowa NJ: Rowman and Littlefield, 1982), pp. 140–146.

[52] W. K. Frankena, "The Concept of Social Justice". In: R. B. Brandt (ed.), *Social Justice* (Englewood Cliffs NJ: Prentice-Hall, 1962), pp. 4–5.

Imagine that a mother gives her son an allowance of 10 dollars each week, her daughter only 5 dollars each week and her neighbor's child no allowance at all. She is clearly treating her neighbor's child more badly, or at least less well, than she is treating her own children. But this does not constitute inequitable treatment because her neighbor's child is not similarly situated in comparison with her own children. He has no moral claim against her to receive an allowance because she is not his parent. Now suppose that her daughter complains that she is being treated unfairly. The mother might explain that both children are expected to buy sports equipment out of their allowances and that her son's golf clubs and grounds fees cost much more than her daughter's tennis racket and court fees. This would show that she is not treating her daughter less well than she is treating her son because she is enabling them equally to meet the expenses of the sport each has freely chosen. Or the mother might explain that her daughter is five years younger than her son and thus not similarly situated within the policy of raising a child's allowance 1 dollar each year. This second consideration would be a justicizing reason, a reason why treating her daughter less well is just. Were the mother to explain that her son is very unruly and that she is giving him a larger allowance to placate him and thus prevent serious disruption within the family, this conceivably might justify treating her daughter worse than she is treating her son, but it would do nothing to show that her action is just or equitable.

For our legal system to deny the constitutional right to procreative autonomy to single adults but confer it upon adults with a partner would presumably constitute inequitable treatment. Single adults are similarly situated with adults in stable relationships regarding procreation because they possess the same human right to procreational autonomy. And to permit the denial or limitation of their liberties to choose whether to procreate and whether to prevent procreation, extremely valuable choices for almost all normal adults, would surely seem to be treating them less well than the law treats those whose procreational choices are protected by the constitutional right to procreational autonomy.

One could, I suppose, argue that this is not so. The fact that single adults have chosen to remain unattached proves that they prefer a life unencumbered by close personal relationships. The law could and should protect their right to remain unencumbered by giving them the right to prevent procreation and to refrain from procreating. But they need not be given the full right to procreational autonomy as long as they remain single. Any adult who really wants a child can at any time freely choose to enter a stable intimate relationship and thereby acquire this constitutional right. Hence, single adults are not really being treated worse than those adults who have chosen a partner and have the constitutional right to procreative autonomy. However, I find this argument unpersuasive. Single adults would be treated worse by this sort of legal policy. While adults in stable relationships would be permitted to choose their lifestyle and at the same time enjoy their human right to procreational autonomy, single adults would be forced either to give up their preferred lifestyle or sacrifice this human right. Thus, this sort of reasoning does not identify a justicizing reason showing that this denial of the constitutional right to procreational autonomy would be equitable.

There might, of course, be some justifying reason that would show this sort of inequitable treatment to be morally acceptable. The obvious consideration is that limiting the procreational choices of single adults would protect potential children from harm. Thomas H. Murray has advanced a line of reasoning that, with modification, might be used to support this view.

> If we are free to avoid having children, shouldn't we have equal freedom to pursue parenthood? The fallacy here is a presumption that the choice to have a child is morally parallel to the choice not to have a child. The former is a choice to initiate a very special human relationship; the latter is a choice to decline such a relationship. The values at the core of the parent–child relationship constrain the former in ways that they do not affect the latter.[53]

He then goes on to identify the family values that are the primary source of meaning in our lives and to suggest that they presuppose stable intimate relationships.

> The values we think of as crucial to family life tend either to be found only in the context of relationships—love, loyalty, affection, trust, care—or, given our social nature, to depend utterly on a foundation of good, enduring relationships—identity, self-confidence, maturation.[54]

From these premises, Murray concludes that

> If children flourish best in stable, loving families, then we harm them by promoting a view of human relationship that equates the decision to initiate such a relationship with the decision to buy a wide-screen television or a medium-priced car. If adults flourish best in enduring, warm relationships and if caring for children also contributes to the flourishing of adults, then we should encourage practices and policies that support such relationships.[55]

Now Murray intends his line of reasoning to show that our society should restrain the unlimited choice of new reproductive technologies because these represent market values in procreation rather than the more appropriate family values. As it stands, his reasoning applies to the ideal scope of the constitutional right to procreational autonomy, not to its possession.

Nevertheless, someone could use his premises to argue that single adults ought to be denied the full right to procreate because any child they might choose to bear and nurture would be harmed. Their child's identity, self-confidence and maturation would be impaired because she would be raised by a parent who has rejected the stable intimate relationship between two parents that normally promotes the central family values of mutual love, loyalty, affection, trust, and care—values found only within close personal relationships.

Let us grant that the most important family values require intimate relationships in order to flourish, perhaps even to exist at all. Nevertheless, single parents usually create and maintain stable intimate relationships with their children. Thus, the family values necessary for the well-being of a child will normally flourish within this parent–child relationship. I conclude that denying the constitutional right to procreational autonomy to single adults while granting it to adults with partners violates their human right to equitable treatment as well as their human right to procreational autonomy. Therefore, single adults ought to possess this valuable constitutional right.

---

[53] T. H. Murray, *The Worth of a Child* (Berkeley CA: University of California Press, 1996), p. 15; see also "What Are Families For? Getting to an Ethics of Reproductive Technology", *Hastings Center Report* 32 (May–June 2002).

[54] *Ibid.*, p. 24.

[55] *Ibid.*, p. 36.

(4) Ought adults on welfare also to possess the same constitutional right to procreational autonomy? Although no one to my knowledge has suggested that being a recipient of welfare is a reason to limit one's right not to procreate or to prevent procreation, there have been and still are those who advocate limiting the legal right to procreate of welfare recipients. Were this done, obviously they would not possess the full right to procreational autonomy. One plausible argument rests on the premise that adults on welfare are unable to care for any future child they might procreate. They would qualify for welfare benefits only if they were poor. This implies that they must lack the economic resources to provide for the needs of their existing family members, whether one or several, much less the needs of any new child. Although there is a general human right to procreational autonomy, including the liberties to procreate and to refrain from preventing procreation, these moral liberties are limited by the duty not to procreate of anyone unwilling or unable to care adequately for a potential child. Hence, adults on welfare do not have any human right to procreate. And this implies in turn that they are not similarly situated with the majority of adults who do possess this human right. Therefore, neither the human right to procreational autonomy nor the human right to equitable treatment requires that the law confer the constitutional right to procreational autonomy upon welfare recipients. On the contrary, they ought to be denied this right in order to protect any future child from being harmed by the inadequate care of her parent or parents.

I doubt that all or even most adults on welfare are unable to care adequately for any future child in some way that distinguishes them, under normal circumstances, from married adults or adults in stable relationships. Although the human right to procreational autonomy is a fundamental right of the individual human being, it takes two to procreate and most parents share the tasks of childcare with a partner. Is a wife who has chosen to abandon her career in order to raise her child able to care adequately for the child's needs without the assistance of her husband? Is either partner, married or unmarried, in a two-career family able alone to provide both financial support and childcare for their child?

In many families, neither partner is able to care adequately for the child without the help of a partner. This is true even though meeting the needs of a child adequately does not require an income sufficient to provide the child with luxuries or parenting skills of the very highest level. If their need of assistance does not imply that they lack the human right to procreational autonomy, why should the fact that adults on welfare require assistance from the public imply that they lack this fundamental moral right? And if it does not, then both their human right to procreational autonomy and their human right to equitable treatment morally require that they possess the constitutional right to procreational autonomy. Moreover, many welfare recipients are better able financially to provide for their children than are the working poor who earn too little to raise themselves above the poverty level yet too much to qualify for Temporary Assistance for Needy Families, Supplemental Security Income, Medicaid, or even food stamps. To deny the constitutional right to procreational autonomy to welfare recipients while granting it to the working poor would presumably be inequitable treatment also.

One way the law could limit the constitutional right to procreational autonomy of welfare recipients would be to terminate the welfare benefits of anyone who gives birth to a child while on welfare. Would this policy violate their human right to privacy? Recall that a private decision is one that affects the life of an individual for better or worse and such that others do not have a sufficient reason to prevent one from acting as one chooses.

Under normal circumstances, the procreational decisions of adults are private decisions such that their human right to privacy implies a moral claim against state interference. Does the human right to privacy of welfare recipients imply the same moral claim against the state that it not interfere with their procreational autonomy? Obviously their decisions whether to procreate affect their lives for better or worse; indeed these are very personal momentous decisions. However, some argue that any decision of a welfare recipient to procreate is not private because it harms the public by increasing the burden imposed upon taxpayers who must fund the welfare system. Thus, their human right to privacy is inapplicable and this limitation on their human right to procreational autonomy is morally justified by the morally compelling state interest in conserving its limited resources.

If having a child increases the welfare payments for which one is eligible or lengthens the time that one will be dependent upon welfare benefits, as is often the case, then any welfare recipient who procreates does thereby impose an additional burden upon society. But is this burden sufficient to imply that his or her decision to procreate is not a private decision? Presumably, one must weight the economic burden imposed upon society against the burden imposed upon the individual denied the right to procreate. Given the momentous nature of most procreational decisions and the fact that any additional burden upon the public welfare system will be only incremental, I would think that the answer is negative. To be sure, when the welfare recipient already has one or more children, the harm imposed by denying the opportunity to have an additional child is also only incremental and usually decreases considerably as the number of children increases.[56] Still, any attempt to specify the number of children after which one loses the right to procreational autonomy seems to be arbitrary and is surely subject to abuse.

Although one might claim that the economic burden upon society is not modest, given the number of welfare recipients who will probably procreate, one must then also multiply the burden upon welfare recipients imposed by denying them procreational autonomy. Thus, multiplying the burdens to be compared will do nothing to change my answer. On the contrary, it makes me more confident that the economic burden on society is not sufficient to justify denying the constitutional right to procreational autonomy of welfare recipients. While this limitation would deny procreational autonomy to all welfare recipients, only those who choose to procreate on welfare would add to the cost of the welfare system. Moreover, now that welfare reform has replaced Aid to Families with Dependent Children with Temporary Assistance for Needy Families, any added burden upon public funds will not continue indefinitely.

One could, as some have, argue that any decision to abuse the public welfare system is not a private decision. At first glance, it does seem unreasonable, and perhaps morally irresponsible, to ask the community to support a child one procreates but for whom one cannot oneself provide financial support. Thus, one might argue that although the decision of a welfare recipient to procreate would momentously affect her or his life for better or worse, it is not a private decision because the state's interest in preventing welfare abuse is morally sufficient to justify interference with this decision.

However, not all decisions of welfare recipients to have a child are morally irresponsible. If one has no children and is nearing the age at which procreation will become

[56] See R. M. Weinberg, "Procreative Justice: A Contractualist Account", *Public Affairs Quarterly* 16 (2002), 417.

impossible or will risk the birth of a seriously defective child or if one knows that one will no longer need or be eligible for welfare benefits in the very near future, then it is not unreasonable to ask society to enable one to enjoy one's human right to procreational autonomy. Under these circumstances, having a child is not a morally abusive act in anything like the way in which lying about one's income in order to receive welfare benefits for which one is not eligible would be to abuse the system. Therefore, the state's interest in preventing welfare abuse is not a morally sufficient reason to justify limiting the right to procreational autonomy of welfare recipients as a class by terminating the welfare benefits of all those who procreate.

Another way the law could choose to limit the constitutional right to procreational autonomy of welfare recipients would be to impose compulsory contraception upon them. Because eligibility for welfare benefits is normally temporary, sterilization would be morally unjustified. However, some have argued that requiring long-term contraception, such as Norplant, ought to be required of all female welfare recipients. (Presumably, equitable treatment would require some comparable form of contraception for male welfare recipients.) However, this would obviously constitute medical treatment without consent and thus violate the welfare recipient's legal right "to determine what shall be done with his own body" as Justice Cardozo held in *Schloendorff*.[57] This right, sometimes called the right to bodily integrity, is a partial protection of the human right to personal security affirmed in Article 3 of the Universal Declaration of Human Rights. This human right is presumably a descendant of the natural right to personal security described more fully by William Blackstone: "The right of personal security consists in a person's legal and uninterrupted enjoyment of his life, his limbs, his body, his health and his reputation."[58] This fundamental human right grounds a moral right of every human being not to be subjected to medical treatment, or any other bodily invasion, without his or her consent. Therefore, any legal regulation that would impose compulsory contraception of welfare recipients would violate their moral right to personal security.

Is there any morally compelling state interest that would justify this violation of a human right? Some have proposed requiring welfare recipients to submit to long-term contraception as a means of helping them to escape from their dependency upon welfare. Experience with welfare programs has shown repeatedly how very difficult it is for recipients to become independent of welfare benefits and how great a social burden this welfare dependency imposes upon society. Some argue that a primary cause of gradually increasing welfare dependency is the birth of additional children who increase the financial needs of the family at the same time that they make it harder for the caregiver to leave her home to earn the necessary income. Given this diagnosis, it would seem that the only way to help welfare mothers to overcome their dependency upon welfare is to require them to submit to compulsory contraception. Arguably, the benefits for welfare recipients and for society as a whole of such a policy would be sufficient to justify it.

Still, experience also shows that many welfare recipients voluntarily refrain from procreating. There is no need to impose compulsory contraception upon them. It is only a minority, probably a relatively small minority, of welfare recipients who are unwilling or unable to refrain from irresponsible procreation. Hence, the legitimate state interest in

---

[57] *Schloendorff v. Society of New York Hospital*, 105 N.E. 92 (1914) at 93.
[58] Sir William Blackstone, *Commentaries*, I, 129, 130.

helping people to overcome their dependency upon welfare would not justify violating the human right to personal security of welfare recipients as a class. Whether it would be justifiable to authorize our courts to require compulsory contraception of a few individual welfare recipients under special circumstances may be left open. At least, we can and should conclude that this line of reasoning is no more persuasive than the others for limiting the possession of the constitutional right to procreational autonomy of welfare recipients in any of the proposed ways.

(5) Ought persons who carry the most serious genetic defects to possess the full constitutional right to procreational autonomy? Although there is a general human right to procreational autonomy, its constituent moral liberties of procreating and of refraining from preventing procreation are limited by one's moral duty not to procreate when one's offspring would probably inherit a defect so serious as to make her life not worth living. Thus, a statute authorizing sterilization of those who are carriers of the most serious genetic defects would presumably not violate the human right to procreational autonomy. This is not to suggest that it might be permissible to sterilize all or even most carriers of genetic defects. It is very likely that every human being carries some genetic defects, and that many persons are carriers of relatively serious defects. It is only those whose genetic code is such that there is a *high* risk that any child they might procreate would inherit a *devastating* genetic disease or defect who might plausibly be denied the full right of procreational autonomy. This narrowly restricted denial of the constitutional right to procreational autonomy would arguably be justified by the morally compelling state interest in preventing the birth of children condemned to endure intolerable suffering while enjoying little or no offsetting value.

Any proposal to authorize the sterilization of carriers of the most serious genetic defects raises the specter of eugenics, a threat that the Supreme Court intended to lay to rest in *Skinner v. Oklahoma*.[59] In that case, Justice Douglas, delivering the opinion of the Court, reasoned that an Oklahoma statute authorizing the sterilization of individuals convicted of three felonies was an unconstitutional violation of the Equal Protection Clause.[60] But if carriers of the most serious genetic defects lack the full human right to procreational autonomy, then they are not similarly situated with the vast majority of citizens who do possess this right and therefore are not in a position to claim that being subjected to compulsory sterilization violates their human right to equitable treatment.

This bodily invasion without their consent would, however, clearly violate their human right to personal security. Is the state interest in preventing the birth of children with lives not worth living, children condemned to endure intolerable suffering, sufficient to justify violating the human right to personal security by sterilizing those who do not possess the full human right to procreational autonomy? It probably would justify this limitation of their constitutional right to procreational autonomy if any such law could promote this state interest efficiently and safely.

However, I doubt that it would be possible to specify those classes of persons to be subjected to compulsory sterilization so as to achieve this goal efficiently. There are too many variables that determine the genetic inheritance of one's offspring. For one thing,

---

[59] *Skinner v. Oklahoma*, 527 U.S. 535 (1942).
[60] *Ibid.* at 541.

the risk of procreating a seriously defective child depends upon the genetic code of one's sexual partner as well as one's own genetic code. And many inherited defects vary in degree from relatively mild to intolerably severe. Moreover, how seriously should one take the risk of inheriting a disposition to become ill or defective later in life, often after many years of happy productive activity? Consequently, legislators would be faced with the dilemma of either defining the classes of those subject to sterilization so broadly that the law would apply to many individuals who could exercise their human right to procreate without bearing or begetting a seriously defective child or specifying those subject to sterilization so narrowly that the law would do little to prevent the birth of children with wrongful lives.

At this point, the safety of enacting any such legislation becomes apparent. Considered as a purely medical procedure, sterilization is relatively safe. Considered as an operation that might well unnecessarily deprive the patient of the opportunity to have children of his or her own, it is far from safe. And considered as a legal rule or policy, it is dangerous indeed. The specter of eugenics does and should haunt us here. Although it is easy enough to discredit the earlier scientific theories that were used to justify similar legislation in the past, it is all too easy to imagine that modern medical science provides a solid basis for eugenics today. But I doubt that this is true for the reasons given in the previous paragraph. Moreover, the judgment that some life is not worth living, a life so different from our own that it is almost unimaginable, is typically unreliable and often prejudicial. Witness the tendency to abort fetuses "suffering" from Down's syndrome even though many persons with this genetic defect live long and reasonably happy lives and even those most severely defective suffer no pain and are unable to recognize how disadvantaged they are. To be sure, they do require extra care and special services that impose psychological and financial burdens upon their families and often upon our public institutions also. However, this fact should be seen as an additional excuse for abuse because of our tendency to exaggerate these psychological burdens and our temptation to reduce the cost of social services even when most needed.

My conclusion is that persons who carry even the most serious genetic defects ought not to be deprived of the possession of the full constitutional right to procreational autonomy even though their human right to procreational autonomy is more limited in scope. Possession of this constitutional right would tend to prevent abuses of their human right to procreational autonomy, limited as it may be, by placing a heavy burden of proof upon those who would restrict their procreational choices.

(6) Ought convicted child abusers to possess the full constitutional right to procreational autonomy? The general human right to procreational autonomy is not unqualified. One who is unwilling or unable to provide adequate care for a child has a moral duty not to procreate.[61] Hence, he or she lacks the moral liberties to procreate and to refrain from preventing procreation. Obviously any parent who abuses his or her child is unwilling or unable to provide adequate care for that child. Therefore, it would seem that a statute authorizing the sterilization of proven child abusers would not violate their human right to procreational autonomy.

---

[61] Compare J. D. Arras and J. Blustein, "Reproductive Responsibility and Long-Term Contraceptives". In: E. H. Moskowitz and B. Jennings (eds.), *Coerced Contraception?* (Washington DC: Georgetown University Press, 1996), pp. 114–116.

Nor would it violate their human right to equitable treatment. Because convicted child abusers lack the human right to procreational autonomy, they are not similarly situated with the vast majority of adults who do possess this right. And the fact that they have engaged in serious child abuse is a just-making reason for treating them worse than other adult human beings. For similar reasons, a statute authorizing the sterilization of convicted child abusers could not be shown to be unconstitutional by the equal protection reasoning of Skinner.

Any such statute would, however, violate their human right to personal security. Would the state interest in preventing the serious harms of child abuse be sufficient to justify such a statute in spite of its invasion of the bodily integrity of child abusers? Well, the prevention of child abuse is surely a morally permissible and probably a morally required goal of state action. And a state policy of sterilizing child abusers would be less subject to abuse than a policy of sterilizing carriers of genetic defects. For one thing, although it is probably impossible to predict reliably who will become a child abuser, the science of psychology probably enables us to predict who will continue to abuse children more reliably than medical science enables us to predict who will procreate a seriously defective child. For another, actions that constitute child abuse can be defined more easily and objectively than can the kinds of lives that are not worth living and thereby constitute wrongful lives.

Nevertheless, the state interest in preventing child abuse, morally compelling as it is, is not sufficient to justify a statute authorizing the sterilization of child abusers. This is because sterilizing child abusers is neither a necessary nor a sufficient means to promote this important state interest. It is not necessary because the courts can prevent an abusive parent from harming his or her child by denying custody and then permitting visitation only under supervision.[62] Erika T. Blum argues that this is not true. She points out that our courts generally allow convicted child abusers to continue to have children and that the many state agencies charged with protecting children from parental abuse are underutilized and understaffed.[63] But the inadequacy of our child protection agencies shows the necessity of increasing their effectiveness rather than evading this problem by compulsory sterilization after serious child abuse has led to a conviction. Indeed, the reliable identification and conviction of child abusers requires this. It is not sufficient because sterilizing a child abuser would do nothing to prevent that person from abusing any of his or her previously born children or, for that matter, children outside the family.[64] In any event, a policy of denying custody and unsupervised visits would be morally preferable because it would not burden the procreative rights of the sexual partner and would deny only the abuser's right to custody, a right forfeited by proven acts of child abuse. My conclusion is that child abusers ought to possess the constitutional right to procreational autonomy, but that should they be proven to have engaged in a pattern of child abuse, the law should limit their access to vulnerable children.

---

[62] For example in *Hanke v. Hanke* 615 A. 2nd 1205 (1992).

[63] E. T. Blum, "When Terminating Parental Rights is not Enough: A New Look at Compulsory Sterilization", *Georgia Law Review* 28 (1994), 1006, 1008.

[64] For a different conclusion, see E. T. Blum, "When Terminating Parental Rights Is Not Enough: A New Look at Compulsory Sterilization", *Georgia Law Review* 28 (1994) 977–1014.

(7) Ought mentally retarded persons to possess the full constitutional right to procreational autonomy?[65] This question is formulated much too generally for illuminating consideration. There is a wide range of mentally retarded persons with various kinds of retardation and to very different degrees. As Justice Pashman observed, delivering the opinion of the Supreme Court of New Jersey,

> Many mentally impaired persons and others with legal disabilities are capable of making their own decisions regarding procreation and sterilization .... We emphasize that there are widely different degrees of mental retardation. The fact that a person is legally incompetent for some purposes ... does not mean that he lacks the capacity to make a decision about sterilization.[66]

Accordingly, it would be a mistake to accept any sweeping assumptions about the moral rights of retarded persons as a class. Probably, some mentally retarded persons possess the full human right to procreational autonomy, although perhaps with less moral weight than normal, others with more limited psychological capacities may possess a limited right to procreational autonomy, and the most severely retarded probably have no human right to procreational autonomy at all.[67] In order to reach some tentative conclusions, I will make a very rough and ready distinction between mildly retarded and severely retarded persons.

Would a state statute authorizing the compulsory sterilization of mildly retarded persons infringe their human right to procreational autonomy? It probably would if it were applied to the majority of such persons. Many persons with limited psychological capacities are able to understand fairly well what difference it would make to their lives whether they procreate, refrain from procreation or prevent procreation. Although they may appreciate the significance of these choices for their lives and for the lives of others less clearly and fully than psychologically normal persons can, this would merely reduce the strength of the grounds of the human right to procreational autonomy, not eliminate them altogether.

Nevertheless, some might try to justify this infringement of the human right to procreational autonomy by appealing to the state interest in preventing the birth of a child who would suffer from mental retardation. Circuit Judge Craven, in the opinion of a United States District Court, accepted this state interest as legitimate and a partial justification of a North Carolina statute that authorized the sterilization of mentally retarded persons.[68] However, he also assumed that the statute would not be applied to the vast majority of retarded persons.

> We hold that the statute is not overly broad. Although it permits initiation of the sterilization procedure against any and all members of the class, it does *not*

---

[65] The similar question about mentally ill persons is raised by the North Carolina statute that authorized the sterilization of "Persons Mentally Ill and Mentally Retarded". See *N. C. Ass'n. For Retarded Children v. State of N. C.* 420 F. Supp. 451 (1976), at 453.

[66] *In Re Grady*, 426 A. 2nd 467 (1981) at 4820483.

[67] See C. Wellman, *Real Rights* (New York and Oxford: Oxford University Press, 1995), 126–131.

[68] *N. C. Ass'n For Retarded Persons*, at 457–458.

> contemplate that all members of the class will be sterilized. Nor is the standard of selection so vague that it cannot be comprehended and applied.[69]

If this interpretation of the statute is correct, then presumably it would seldom if ever be applied to an individual who is only mildly retarded. Still, I am reluctant to accept his assumption that the state interest in preventing the birth of a retarded child is ever sufficient to justify the compulsory sterilization of a retarded person. It is too difficult to predict when a retarded person would give birth to a seriously retarded child, and the danger of abusing any such statute is too great.[70]

The other state interest North Carolina specified to justify its statute was the state interest in preventing the birth of a child that the retarded parent would be incapable of caring for adequately. This is a much more plausible justification because one can predict much more reliably when a person lacks the psychological capacities required for even minimally acceptable childcare than when one's genetic code would result in the birth of a seriously defective child. However, as Judge Craven recognized, this would almost never be true of persons who are only mildly retarded. "Secondly, the statute presumes that some persons may be so severely retarded that they would be unable to properly care for a child should they conceive one."[71] When this is the case, compulsory sterilization would not violate the individual's human right to procreational autonomy because his or her moral duty not to procreate limits the general liberty of procreation. My conclusion is that, unless drawn very narrowly indeed, any statute that would authorize the sterilization of persons who are only mildly retarded would be an unjustified violation of their human right to procreational autonomy.

Would a state statute authorizing the compulsory sterilization of severely retarded persons violate their human right to procreational autonomy? I believe that in most cases it would not do so simply because most severely retarded persons do not possess any human right to procreational autonomy. As Justice Pashman recognized

> What is at stake is not simply a right to obtain contraception or to attempt procreation. Implicit in both these complementary liberties is the right to make a meaningful choice between them. Yet because of her severe mental impairment, Lee Ann does not have the ability to make a choice between sterilization and procreation, or between sterilization and other methods of contraception ....[72]

Because most severely retarded persons are incapable of making meaningful choices regarding either procreation or the prevention of procreation, it is idle and misleading to ascribe any human right of procreational autonomy to them. Moral rights, as best understood, concern who ought to possess dominion, freedom and control, over some defining core in some possible confrontation with one or more second parties. Therefore, only individuals capable of exercising freedom and control can possess moral rights, and those with limited capacities can possess moral rights only where their capacities for agency are sufficient to enable them to make meaningful choices.[73] It follows

---

[69] *Ibid.* at 458, italics in original.
[70] See *In Re Grady*, at 472.
[71] *Ibid.* at 456.
[72] *Ibid.* at 474.
[73] See C. Wellman, *Real Rights*, pp. 128–131.

that most severely retarded persons have no fundamental moral right to procreational autonomy.

Some of them could, however, possess a human right to personal security. This might not be the full human right to personal security possessed by normal adults because severely retarded persons probably cannot understand many of the threats to their persons or the means available to protect themselves. But presumably many severely retarded persons are capable of making meaningful choices between being beaten or stabbed and not being harmed in those ways. However, almost all severely retarded persons are incapable of making informed choices concerning any but the most simple medical treatments. Presumably, those who are severely retarded are incapable of deciding whether sterilization is in their best interests. Therefore, their human right to personal security would not be violated by being sterilized without their consent.

Even granted that severely retarded persons have none of the human rights normally protected by the constitutional right to procreational autonomy, this does not imply that a state statute authorizing their sterilization would be morally justified. Because the state's moral duty to protect its citizens from harm applies most strongly to those who are most vulnerable, compulsory sterilization would be justified only when it is in the best interests of those subjected to it. Could sterilization ever pass this test? I believe that in the case of the most severely retarded persons it could and often does. For example, consider the plight of Lee Ann Grady.

> As Lee Ann has approached the age of 20—when she will leave her special class in the public school system—the Gradys have given more thought to her future. The parents fear they will predecease their daughter and she will be unable to live independently. Thus, they have sought to attain for her a life less dependent on her family. The Gradys wish to place Lee Ann in a sheltered work group and eventually in a group home for retarded adults. But the parents see dependable and continuous contraception as a prerequisite to any such change in their daughter's environment.[74]

Or consider Valerie N.

> An incompetent developmentally disabled woman has no less interest in a satisfying or fulfilling life free from the burdens of an unwanted pregnancy than does her competent sister . . . . If the state withholds from her the only safe and reliable method of contraception suitable to her condition, it necessarily limits her opportunity for habilitation and thereby her freedom to pursue a fulfilling life.[75]

To be sure, there are very many other cases in which the state's duty is to protect mildly retarded persons from compulsory sterilization. But a statute authorizing the sterilization of severely retarded persons when it is in their best interests would be morally justified.

The question remains as to how the law could best authorize the sterilization of the severely retarded while at the same time preventing the wrongful sterilization of those who are only mildly retarded. The obvious solution would be to limit the possession of

---

[74] *In Re Grady*, at 470.

[75] *Conservatorship of Valerie N.*, 707 P.2nd 760 (1985) at 773.

the constitutional right to procreational autonomy to mildly retarded persons and to deny its possession to those who are severely retarded. The problem is that this would require drawing the line between mildly retarded persons and severely retarded persons in very general terms. I doubt that this could be done in a way that would distinguish between cases in which compulsory sterilization would not be justified and cases in which it would be in the best interests of those subjected to it. There are simply too many kinds of mental retardation, present in greater or lesser degrees and suffered by persons living under a wide variety of circumstances.

Our courts have chosen a different legal solution to this problem. They have assumed that all mentally retarded persons possess the constitutional right to procreational autonomy, but specified procedures by which a court, usually acting on the petition of a parent or guardian, has the authority to permit or order the sterilization of some severely retarded individual. Thus, Justice Pashman argued

> The right to choose among procreation, sterilization and other methods of contraception is an important privacy right of all individuals. Our courts must preserve that right. Where an incompetent person lacks the mental capacity to make that choice, a court should ensure the exercise of that right on behalf of the incompetent in a manner that reflects his or her best interests.[76]

Similarly, Justice Grodin reasoned, "True protection of procreative choice can be accomplished only if the state permits the court-supervised substituted judgment of the conservator to be exercised on behalf of the conservatee who is unable to personally exercise this right."[77] By assuming that all retarded persons possess the constitutional right to procreative autonomy, the courts hope to minimize the abuse of the practice of sterilization by placing a heavy burden of proof upon those who would impose it upon retarded persons incapable of giving their consent. Yet by specifying procedures by which a court could authorize the sterilization of a severely retarded person, our courts have sought to protect the best interests of those few individuals for whom this would be justified.

Not everyone is satisfied with this solution. Justice Lucan, concurring and dissenting in *Conservatorship of Valerie N.*, argued

> I find fundamentally problematic my colleagues' conclusion that there is a constitutional right to "substituted consent" in this context . . . I worry whether the "rights" which we are "protecting" are in fact more likely to become those of the incompetent's caretaker.[78]

Chief Justice Bird, dissenting, objected even more strongly

> Today's holding will permit the state, through the legal fiction of substituted consent, to deprive many women permanently of the right to conceive and bear children . . . . The majority opinion opens the door to abusive sterilization practices which will serve the convenience of conservators, parents, and service providers rather than incompetent conservatees. The ugly history of

[76] *In Re Grady* at 475.
[77] Conservatorship of Valerie N. at 777.
[78] *Ibid.* at 779.

> sterilization abuse against developmentally disabled persons in the name of seemingly enlightened social policies counsels a different choice.[79]

I agree that the doctrine of substituted judgment is a legal fiction. The consent or request of a parent, guardian, or conservator cannot literally be said to substitute for the consent or request of an individual completely incapable of consenting or requesting. But legal fictions are sometimes both useful and legitimate in the law. The crucial question in this context is whether this particular legal fiction can serve the best interests of those who may justifiably be sterilized while at the same time protecting the best interests and moral rights of those who ought not to be subjected to compulsory sterilization.

Justice Pashman argued that the legal fiction of substituted judgment can serve both purposes.

> We do not pretend that the choice of her parents, her guardian *ad litem*, or a court is her own choice. But it is a genuine choice nevertheless—one designed to further the same interests she might pursue had she the ability to decide herself. We believe that having the choice made in her behalf produces a more just and compassionate result than leaving Lee Ann with no way of exercising a constitutional right.[80]

But can the doctrine of substituted judgment concerning one's reproductive capacities be authorized without inviting intolerable abuses? Very mindful of the fact that "sterilization has a sordid past in this country,"[81] Pashman argued that strict procedural safeguards can and would protect only mildly retarded persons from being sterilized.

> The question of who besides the parents has standing to represent the purported interests of the incompetent can await further determination. Nevertheless, we believe that an appropriate court must make the final determination whether consent to sterilization should be given on behalf of an incompetent individual. It must be the court's judgment, and not just the parents' good faith decision, that substitutes for the incompetent's consent.[82]

He then went on to specify the procedures a court must follow in order to authorize the sterilization of a mentally retarded individual. These include:

> Fourth, the trial court must be persuaded by *clear and convincing* proof that sterilization is in the incompetent person's best interests. To determine those interests, the court should consider at least the following [nine] factors .... These factors should each be given appropriate weight as the particular circumstances dictate.[83]

Although it may well be possible to improve upon the procedures specified by Justice Pashman, I believe that in principle he has described the best solution to the legal problem

---

[79] *Ibid.* at 781–782.

[80] *In Re Grady* at 481.

[81] *Ibid.* at 472. See also P. A. Lombardo, "Three Generations, No Imbeciles: New Light on Buckv. Bell", *New York University Law Review* 60 (1985), 30–62.

[82] *Ibid.* at 475.

[83] *Ibid.* at 483, italics in original.

of authorizing the sterilization of severely retarded persons when it is in their best interests while protecting the moral right of only mildly retarded persons not to be subjected to sterilization without their consent. All mentally retarded persons ought to possess the constitutional right to procreational autonomy, but the courts ought to be authorized to permit or order sterilization provided that they follow procedures designed to protect the moral rights and personal interests of all retarded individuals. Although it is impossible justifiably to limit the possession of the constitutional right to procreational autonomy in general terms, it is practicable to specify the criteria a court ought to apply when deciding whether some retarded individual may justifiably be sterilized.

(8) Ought minors, persons under the age of legal competence, to have the full constitutional right to procreational autonomy? Two limitations of this right are plausible enough to reward consideration. There have been a number of court cases concerning State statutes that restrict the decision of a minor to prevent procreation by requiring parental consent before a physician may perform an abortion.[84] And there have been proposals to limit the minor's right to procreate by requiring adolescents to submit to the implantation of Norplant. Are limitations such as these morally justified?

If minors possess the full human right to procreational autonomy, then limitations of their analogous constitutional right would be suspect. But do they possess this fundamental moral right? Although some do, others have only a partial human right to procreational autonomy and some have no such right at all. Let me explain. Although children are certainly human beings, they are not born with any human rights. Because rights concern the proper allocation of dominion, freedom, and control, only moral agents, beings capable of exercising freedom and control in the light of moral reasons, are possible possessors of moral rights. This implies that infants, who lack the psychological capacities for moral agency, cannot be said to possess any rights. However, as children grow and develop their capacities for rational choice, they gradually acquire moral rights. Presumably, they acquire the most simple rights before more complex ones, and each right will be acquired gradually depending upon the complexity of its various parts.[85] What follows from this is that very immature minors possess no human right to procreational autonomy and that other minors acquire this fundamental moral right gradually as they become more and more mature. Therefore, I applaud the opinion of Justice Blackmun that

> We agree with applicants and with the courts whose decisions have just been cited that the State may not impose a blanket provision . . . requiring the consent of a parent or person *in loco parentis* as a condition for abortion of an unmarried minor during the first 12 weeks of her pregnancy . . . . Constitutional rights do not mature and come into being magically only when one attains the state-defined age of majority.[86]

Presumably, the same reasoning applies to any blanket State requirement that would subject minors to the implantation of Norplant without their consent.

---

[84] For example, *Planned Parenthood of Missouri v. Danforth*, 428 U.S. 52 (1976), and *Bellotti v. Baird* 443 U.S. 622 (1979).

[85] See C. Wellman, *An Approach to Rights*, pp. 139–140.

[86] *Danforth* at 74, italics in original.

If my conception of moral rights is correct, then minors will begin to acquire the human right to procreational autonomy at different ages and the scope of the right each has acquired will cover different procreational decisions at different times in their lives. This is because the relevant considerations are diverse and require different kinds of psychological capacities and background experiences to be fully appreciated. For example, as Justice Powell observed:

> Particularly when a minor becomes pregnant and considers an abortion, the relevant circumstances may vary widely depending upon her age, maturity, mental and physical condition, the stability of her home if she is not emancipated, her relationship with her parents, and the like.[87]

This extreme variation will make it very difficult indeed to determine in any given case whether some legal limitation on a specific procreational decision of a minor infringes his or her human right to procreational autonomy. Would it be permissible for a state to make the administration of the law easier and at the same time to increase the legal certainty of procreational law by setting an age limit for the possession of the constitutional right to procreational autonomy?[88] I think not. The variability of the morally relevant considerations that tempts one to simplify the law in this way renders it morally impermissible. If, and that remains undecided, the law may permissibly limit the constitutional right to procreational autonomy of minors in any way, it ought to be done by authorizing a court to determine in each case whether the minor involved has the moral maturity to make the specific decision at issue.

If the minor is in fact too immature to make a rational informed decision concerning the procreational issue at hand, then the law could deny him or her that measure of autonomy without infringing the human right to procreational autonomy.[89] If, on the other hand, the minor is morally mature enough to make that choice, then the State interest in protecting the minor, vulnerable as she or he may be, presumably would not justify any such limitation.[90] Finally, if the minor is unwilling or unable to care for a child adequately, then the law could impose long-term contraception, such as Norplant, upon the minor without infringing her human right of procreational autonomy because the general moral liberty to procreate is limited by one's inability to care for one's child. Such cases would probably be fairly frequent because many teenage mothers never escape the burdens of poverty and lack of education, and these effects are often passed on to their children.[91]

Any such limitation might, however, infringe some other human right. As a child's ability to make rational choices develops, she acquires a human right to privacy with a gradually increasing scope.[92] This is because the reasons to respect the moral agency of a human being then apply and imply a moral claim against others that they not prevent or hinder one from acting as one has chosen unless they have a moral reason sufficient to justify such interference. Hence, it would seem that any State interference with a

---

[87] *H. L. v. Matheson*, 450 U.S. 398 (1981), at 419.
[88] See the opinion of Justice Stevens in *Danforth*, at 104–105.
[89] Compare *Carey v. Population Services International*, 431 U.S. 678 (1977) at 705.
[90] Compare *Bellotti*, at 650.
[91] Blank and Merrick, p. 80.
[92] Compare *Carey*, at 693.

mature minor's procreational decisions would infringe the human right to privacy. But is the decision of an unemancipated minor a private decision? Clearly the decision of a dependent minor to procreate or, if pregnant, whether to have an abortion will have very serious consequences for her parent or parents. Would the burdens that any such decision would impose upon a parent constitute a sufficient reason to limit the scope of the morally mature minor's fundamental moral right to privacy? Again, the morally relevant circumstances will vary greatly from case to case. How young is the dependent minor? How much of her education has she completed and how soon will she be able to earn an income sufficient to support herself and her child? Is she willing to put her child up for adoption? How strong are her religious and moral convictions concerning abortion? Are her parents understanding and supportive or dogmatic and dictatorial? Once more, I believe that even unemancipated minors ought to possess the constitutional right to procreational autonomy limited only by the power of a parent to petition a court to overrule the procreational choice of the minor when this is justified by the State's interest in maintaining parental discretion regarding the care of a child.[93]

The parents are not the only parties who might be seriously burdened by the procreational decision of a minor. The cost to the public of teenage childbearing amounts to several billion dollars each year.[94] Does this fact imply that any procreational decision by an adolescent is not a private decision? Obviously it does not imply this regarding the decision of a pregnant minor to have an abortion, but it might show that in some cases a decision to procreate or to refrain from preventing procreation is not a private decision. However, this would be true only when neither the minor nor her family are willing and able to care for her child and the financial burden upon some State welfare program or social service will be unusually severe. Even then, the only morally justified limitation of the minor's constitutional right would be to authorize a court to require her to submit to Norplant or some comparable long-term contraceptive that would merely delay her ability to exercise her fundamental moral right to procreate.

Nevertheless, to require a mature minor to submit to the implantation of Norplant without her consent presumably would be to violate her human right of personal security. As John Robertson argued

> Teenagers do have a strong interest in bodily integrity. The insertion of the device may be viewed as minor, but the potential side effects are serious enough—many women cannot tolerate Norplant—to make the bodily intrusion substantial. Overriding this interest may be difficult to justify, despite the worthiness of the goal.[95]

Thus, the mature minor's human right of personal security is an additional moral reason against this sort of limitation of her constitutional right to procreational autonomy. And it may well apply in some cases where her human right to procreational autonomy does not. This is because many decisions whether to consent to or refuse medical treatment are probably less complex and difficult than typical procreational choices. Hence,

---

[93] Compare *Bellotti*, at 637.
[94] Blank and Merrick, p. 80.
[95] *Children of Choice*, p. 92.

minors probably acquire this part of the human right to personal security earlier and more completely than the human right to procreational autonomy.

On the other hand, if Robertson has described what is involved regarding personal security in this sort of case, then the weight of this right is less than the usual weight of the right to procreational autonomy. Therefore, the State's interests in reducing the burden on the funding of public programs together with its interest in maintaining parental discretion within the family may sometimes be sufficient to justify violating this human right. Of course, the State's moral duty to protect the most vulnerable members of society, including mature minors, must be considered also. In some cases her best interests will count against respecting the decision of a mature minor, and in others it will support her decision to refuse to submit to Norplant. Accordingly, the law ought to permit this violation of a mature minor's human right of personal security only when a court has balanced all the relevant circumstances and ordered the implantation of Norplant.

Although, I have argued that all minors mature enough to possess the human right to procreational autonomy ought to possess the constitutional right to procreational autonomy, I have conceded that our courts ought to be permitted to deny the exercise of this right under very special circumstances. But in practice, our courts might well impose these limitations in a discriminatory manner. As John Robertson points out, there is a serious problem in identifying the target of any policy of reducing teenage pregnancy by requiring Norplant. "If directed at all female adolescents, it would be grossly overbroad, intruding upon the many to prevent pregnancy by a few. If targeted to subgroups that have high rates of pregnancy, it risks actual or perceived discrimination on racial or ethnic grounds."[96] The statistics concerning teenage pregnancy indicate that the impact of any program of fertility control using Norplant will be greatest upon black minors, especially those who depend upon public assistance.[97] This suggests that it would involve racial discrimination and the inequitable treatment of those in poverty. Finally, any effective statute intended to reduce irresponsible pregnancies of minors must apply primarily, perhaps exclusively, to young women. Hence, it might necessarily constitute inequitable treatment of females.

In short, even my modest suggestion that the constitutional right to procreational autonomy of minors may occasionally be limited in some individual cases under special circumstances threatens the human right to equitable treatment of a significant number of mature minors. The defining core of this fundamental moral right is one's moral claim not to be treated worse than those who are comparably situated unless there is some justicizing or just-making reason for such treatment. In the cases with which we are concerned here, all those who possess the human right to procreational autonomy are comparably situated. Now the State's interests that might justify limiting the exercise of a mature minor's constitutional right to procreational autonomy in the cases I have proposed are not justicizing reasons. That is to say that they neither show that the mature minor lacks the human right to procreational autonomy and thus is not similarly situated with those who do possess it nor do they show that limiting her exercise of this right is not unfair to her. Nevertheless, if they involve very important personal or social values, they may be sufficient to morally justify violating the minor's human right to equitable treatment. The problem that remains is the danger that in practice the courts will impose Norplant upon

[96] *Ibid.*
[97] Blank and Merrick, p. 82.

individuals in cases where this is not morally justified. To prevent this abuse, the courts should attempt to specify the considerations that must be taken into account whenever a court does limit the procreational autonomy of a mature minor.[98] Whether it will be possible to eliminate, or at least greatly reduce, inequitable treatment in this way only time can tell.

Everything considered, ought minors, persons under the legal age of competence, to possess the constitutional right to procreational autonomy? Because a primary purpose, perhaps the primary purpose, of marriage is to legitimize procreation and confer the right to found a family, married minors ought to have the same procreational rights as married adults. Of course, at what age or under what conditions minors ought to possess the constitutional right to marry is a question too difficult to be settled here. Presumably, emancipated minors also ought to possess the same procreational rights as adults. However, as long as a minor remains dependent upon his or her parents, their interests and the State's interest in supporting their parental discretion regarding the care of their child are factors that need to be considered in any case involving a questionable procreational choice.

What remains is the controversial question of how, if at all, procreational law ought to recognize the distinction between moral maturity and moral immaturity. This is a complicated, even a messy, problem because different individuals achieve moral maturity, the capacity to understand and act on moral reasons, at different ages and more quickly concerning some kinds of choices than others. Therefore, minors will acquire the human rights to procreational autonomy, privacy, personal security, and equitable treatment more or less quickly and to differing degrees as their abilities to make informed rational choices in each of these domains grow. My conclusion is that it would be a mistake for the law to attempt to draw any clean line between minors who do possess and those who do not possess the right to procreational autonomy. It ought to confer this constitutional right upon all minors. However, it ought also to authorize courts to limit the exercise of this right by a few individual minors under special circumstances. And in such cases, the State ought to have a very heavy burden of proof that any such limitation is necessary and the court should be required to consider a specified variety of legally and morally relevant facts.

9.4. Conclusions

The constitutional right to procreational autonomy is a rights-package consisting of two distinct but related rights, the liberty-right to procreate or refrain from procreating and the liberty-right to prevent or not prevent procreation. In addition to its defining bilateral liberty, each of these includes a number of associated legal positions, especially a claim against interference with the right-holder's exercise of his or her constitutionally protected liberty.

Who does in fact possess this right under existing constitutional law? The line of cases from *Meyer* and *Griswold*, merging in *Eisenstadt* and continuing through *Carey* and *Davis* clearly establishes that most adult married persons possess the constitutional right to procreational autonomy. Presumably, *Eisenstadt* also shows that many, although

[98] Much as Justice Pashman did regarding mentally retarded persons in *In Re Grady* at 483.

not necessarily all, unmarried adults possess the same right. In addition, the reasoning of our courts suggests that men and women equally possess this general constitutional right, although their possession of the more specific procreational rights this implies may differ. Finally, women on welfare, mentally retarded persons, minors, and criminals also possess this right, although their exercise of it may be denied in some circumstances.

Because the holdings in several of these precedents are controversial and the implications of the judicial reasoning in most of them remain rather indeterminate, it is very likely that in the near future there will be cases in which new attempts by the States to limit the procreational autonomy of several classes of persons will be at issue. I have left the prediction of how the courts will decide these cases to specialists in constitutional law. As a moral philosopher, I have asked only what these potential judicial decisions ought to be.

Who ought to possess the constitutional right to procreational autonomy? Married adults ought to possess this right in order to protect their human right to procreational autonomy and to respect their human right to privacy. In addition, (1) unmarried adults in stable heterosexual relationships or in (2) stable same-sex relationships ought to possess the constitutional right to procreational autonomy for the same reason. In neither case is the State's interest in promoting family values or its interest in protecting any offspring from harm sufficient to justify limiting the procreational rights of unmarried couples. (3) Single adults, adults not living in any stable relationship with a partner, ought also to possess this right to protect their human right to procreational autonomy and to respect their right to equitable treatment. There is no convincing evidence that they are in general incapable of providing adequate care for their children and no State interest sufficient to justify treating them worse than married persons who are similarly situated as possessors of the human right to procreational autonomy. (4) Although some have proposed terminating the welfare benefits of women who have an additional child, this would violate their human rights to procreational autonomy and to privacy; and imposing compulsory contraception upon them would also violate their human right to personal security. Because no State interest is sufficient to justify any such violations, welfare recipients ought to possess the constitutional right to procreational autonomy. (5) Although persons who carry the most serious genetic defects do not possess the full human right to procreational autonomy, they ought to possess the constitutional right to procreational autonomy because any State statute that would limit their right to procreate would invite intolerable abuse. (6) Again, child abusers do not possess the full human right to procreational autonomy, but they ought not to be denied the analogous constitutional right in order to protect them from unjustified sterilization. Compulsory sterilization is neither necessary nor sufficient to promote the State interest in preventing child abuse. Instead, child abusers ought to be denied unlimited access to children. Because mentally retarded persons and minors vary so greatly in their capacities for making informed rational decisions, some do and others do not possess the relevant human rights to procreational autonomy, privacy, personal security, or equitable treatment. Therefore, it would be misguided to attempt to draw any clear line between those who do and those who do not possess the constitutional right to procreational autonomy. Constitutional law ought to confer this right upon (7) all mentally retarded persons and (8) all minors, but authorize the courts to limit their exercise of this right under very exceptional circumstances when there is clear and convincing evidence that the relevant considerations justify the specific limitation at issue in the individual case.

I do not pretend that any of my conclusions is unquestionably correct or that any of the arguments I have given for them is conclusive. What I do believe is that I have shown how several fundamental human rights are morally relevant to the possession of the constitutional right to procreational autonomy and how difficult it would be to justify infringing these rights by any appeal to legitimate State interests or other morally relevant considerations. In this way, I hope to have contributed to a more enlightened and reasonable consideration of some of the urgent legal issues that will confront our courts in the near future.

CHAPTER 10

MEDICAL FUTILITY AND MORAL RIGHTS

Patients, or their families, sometimes insistently request medical treatments that their physicians believe would be either ineffective or of no benefit. It is not uncommon for someone suffering from an ordinary cold to request that his physician prescribe an antibiotic even though the physician knows that antibiotics are effective only against bacteria, not viruses. A mother might demand that her newborn child be given intensive neonatal care even though he is so premature that he has almost no chance of survival. A patient terminally ill with cancer might beg for a third series of chemotherapy treatments even though the most optimistic prognosis is that this would merely postpone death for a very few weeks of intractable pain. A husband may insist that his wife be continued on intensive care even after she has lapsed into an irreversible coma or persistent vegetative state. Ought physicians, and hospitals, to be legally permitted to refuse to provide futile medical care demanded by patients, or their families? This is an urgent question that arises in current medical practice, has occasioned lively debate in legal periodicals and philosophical journals, and has produced a number of court cases.

10.1. Recent Futility Cases

In re Wanglie.[1] On January 1, 1990, Mrs. Helga M. Wanglie, an 85-year-old woman, was taken from a nursing home to Hennepin County Medical Center for emergency treatment requiring intubation and artificial respiration. In May, her heart stopped beating during an attempt to wean her from the respirator, and in early June, physicians concluded that she was in a persistent vegetative state. In June and July, physicians suggested that life-sustaining treatment be withdrawn, but her husband, daughter, and son insisted on continued treatment. Eventually, the hospital petitioned a Minnesota District Court to appoint an independent conservator to decide whether the respirator was beneficial to the patient. However, the District Court denied the petition and held that Oliver Wanglie was the best person to be the guardian for his wife.[2]

In re Jane Doe.[3] Jane Doe, a 13-year-old child who had experienced medical problems since birth was admitted in May 1991 to Scottish Rite Hospital following a choking episode and over the next weeks her physical state deteriorated until her condition varied

[1] *In Re the Conservatorship of Helga M. Wanglie*, Fourth District Court, Hennepin County Minnesota, Probate Court Division, No. PX-91-283, 1991.

[2] S. H. Miles, M.D., "Informed Demand for 'Non-Beneficial' Medical Treatment", *New England Journal of Medicine* 325 (1991), 512–513.

[3] *In re Jane Doe,a minor*, 418 S.E.2nd 3 (Ga. 1992).

between stupor and coma. In late May, her doctors placed her on a respirator and later inserted breathing and feeding tubes. In August and again in September, when her condition continued to decline, her doctors suggested de-escalation of life support and entering a Do Not Resuscitate (DNR) order. Although her mother consented to a DNR order, her father refused. The hospital then filed a declaratory judgment action seeking guidance as to which of the parent's wishes it should follow.[4] Upon appeal, the Supreme Court of Georgia ruled that because the father had revoked the consent of the mother, the hospital could not enter a DNR order.[5]

In the Matter of BABY "K."[6] Baby K was born at the hospital in October 1992 with anencephaly, a congenital malformation in which a major portion of the brain, skull, and scalp are missing. When Baby K had difficulty breathing, physicians placed her on a mechanical ventilator. The physicians explained to her mother that most anencephalic infants die within a few days of birth. Because aggressive treatment would serve no therapeutic or palliative purpose, they recommended that Baby K only be provided with nutrition, hydration, and warmth. Ms. H insisted that Baby K be provided with mechanical breathing assistance whenever the infant developed difficulty breathing on her own. Following Baby K's second admission, the hospital filed a declaratory action to resolve the issue of whether it was obligated to provide emergency medical treatment to Baby K that it deemed medically and ethically inappropriate.[7] The United States District Court, E. D. Virginia, considered a wide range of arguments advanced by both parties and finally denied the hospital's petition.[8] However, the United States Court of Appeals, Fourth Circuit, narrowed the issue as to whether the federal Emergency Medical Treatment and Active Labor Act (EMTALA) required that the hospital provide treatment other than warmth, nutrition, and hydration to Baby K.[9] Interpreting the EMTALA in accordance with its plain language, it held that the hospital was legally required to provide emergency stabilizing treatment including mechanical respiration in the event that Baby K were to be readmitted for respiratory distress in the future.[10]

Bryan v. Rectors and Visitors of the University of Virginia.[11] On February 5, 1993, Shirley Robertson was transferred to the University of Virginia Medical Center for emergency treatment. After her respiratory distress was stabilized, she was admitted to the hospital for further medical treatment. Although the hospital received instructions from Mrs. Robertson's husband and all of her children to take all necessary measures to keep her alive and trust in God's wisdom, on February 17 the hospital entered a DNR order. Eight days later, the patient died of heart attack.[12] Cindy Bryan, acting as administratrix of the estate of Shirley Robertson, sued the University of Virginia claiming that its Medical Center had violated the EMTALA by failing to resuscitate Mrs. Robertson when her heart

---

[4] *Ibid.*, at 4–5.
[5] *Ibid.*, at 7.
[6] *In the Matter of BABY "K"*, 16 F.3rd 590 (4th Cir. 1994).
[7] *Ibid.*, at 592–593.
[8] *In the Matter of BABY K*, 832 F.Supp. 1022 (E.D.Va. 1993).
[9] *In the Matter of BABY "K"*, at 592.
[10] *Ibid.*, at 598.
[11] *Bryan v. Rectors and Visitors of the University of Virginia*, 95 F.3rd 349 (4th Cir. 1996).
[12] *Ibid.*, at 350–351.

required stabilizing treatment.[13] The United States Court of Appeals, Fourth Circuit, dismissed her claim on the grounds that the EMTALA requires only stabilizing treatment for a patient's emergency condition, a requirement concededly met by the Medical Center. Any further treatment would be governed, not by the EMTALA, but by the state medical malpractice law.[14]

Causey v. St. Francis Medical Center.[15] Having suffered cardiorespiratory arrest, Sonya Causey, who was comatose, quadriplegic, and in end-stage renal failure was transferred from a nursing home to Saint Francis Medical Center. Although her physician believed that continuing dialysis would have no benefit, the patient's family demanded aggressive life-sustaining care. After the hospital's Morals and Ethics board agreed with her physician's opinion to discontinue dialysis and life-support procedures and to enter a DNR order, the patient was taken off a feeding tube and other similar devices. The day the ventilator was removed, Sonya Causey died. The husband, father, and mother of Sonya Causey then petitioned for damages against the Saint Francis Medical Center and her physician.[16] The Court of Appeal of Louisiana, Second Circuit, affirmed the judgment of the trial court dismissing the plaintiffs' action as premature because it was brought as a medical malpractice tort action and the Medical Malpractice Act requires that the matter should first be submitted to a medical review panel.[17]

Obviously, these cases have done very little to decide what the law actually requires concerning medically futile treatment and even less about how the law ought to develop as new cases arise. The holdings of *Wanglie*, *Jane Doe*, and *Causey* all concerned only procedural issues. And although *Baby K* did deal with a substantive issue, the Court of Appeals considered only one of many legally relevant sources, the EMTALA, and its interpretation of that statute has been seriously questioned, both in the dissenting opinion of Judge Sprouse[18] and in various legal periodicals.[19] Moreover, Bryan restricted the applicability of *Baby K* to emergency treatment and reasoned that all other medical treatment is governed by the rather indeterminate standard of care in medical malpractice law. Accordingly, the question of whether physicians and hospitals ought to be legally permitted to refuse to provide futile medical treatment demanded by patients or their families is as timely as it is urgent for both physicians and their patients.

10.2. Reasons for a Legal Liberty

If legislatures were to enact or the courts to recognize a legal liberty of physicians (and hospitals) to refuse to provide futile medical treatment demanded by patients or their families, it would be necessary to define "futile medical treatment" with some precision.

---

[13] *Ibid.*, at 349–350.
[14] *Ibid.*, at 351 and 353.
[15] *Causey v. St. Francis Medical Center*, 719 So.2nd 1072 (La.App. 2 Cir. 1998).
[16] *Ibid.*, at 1073–1074.
[17] *Ibid.*, at 1072 and 1076.
[18] *In the Matter of BABY "K"* at 598–599.
[19] For example, J. F. Daar, "Medical Futility and Implications for Physician Autonomy", *American Journal of Law & Medicine* 21 (1995), 221, 227–228; M. Strasser, "The Futility of Futility: On Life, Death, and Reasoned Public Policy", *Maryland Law Review* 57 (1998), 505, 508–509.

Although some knowledgeable persons doubt that this is possible,[20] there is no doubt that the concept of medical futility plays an important role in contemporary medical practice.[21] Therefore, let us begin with a rough-and-ready definition in terms of ordinary language. The *Oxford English Dictionary* reports that "futile" means "incapable of producing any result; useless, vain" and "occupied with worthless or trivial matters: frivolous, lacking in purpose." Presumably, then, futile treatment would be either incapable of or very unlikely to produce any physiological effect or, although capable of producing some physiological effect, no worthwhile effect. Hence, the literature typically distinguishes between physiological or quantitative futility and qualitative futility.[22]

There are a number of reasons one might well advance for why there ought to be a legal liberty of the physician to refuse to provide futile medical treatment even when it is demanded by the patient or her family. (1) When the demanded medical treatment is incapable of producing any physiological effect, it could not possibly serve any medical purpose and the demand for it is irrational and may be disregarded. But are the purposes of medical treatment limited to producing physiological effects? The Hastings Center *Guidelines* note that "Treatment that is physiologically futile may offer psychological benefits and so may be warranted."[23] It is hard to imagine why medical treatment that could benefit the patient, whether physiologically or psychologically, would serve no medical purpose. Presumably, Rosemarie Tong is correct when she asserts, "Certainly, absent some basic notion of what the proper goals of medicine are (an essentialist concern, to be sure), physicians could not deny patients treatment on the grounds that it is *medically inappropriate*."[24]

Moreover, even if some demanded treatment were medically futile because it would serve none of the defining goals of medicine, it still might not be entirely futile because it might serve some other purpose.

> We can agree that physicians need not administer pointless care, but an intervention can only be called pointless relative to some goal. Occasionally aggressive interventions are "futile" toward any medical goal, as for example, resuscitation efforts on a patient who is clearly dead. But where the intervention will have at least some effect, even if only to prolong the patient's life a few minutes or to appease a family's guilt over their earlier neglect of the patient, the immediate question becomes whether the goal is worth the effort and the pain.[25]

---

[20] P. R. Helft et al., "The Rise and Fall of the Futility Movement", *The New England Journal of Medicine* 343 (2000), 293–294; B. A. Brody and A. Halevy, "Is Futility a Futile Concept?", *The Journal of Medicine and Philosophy* 20 (1995), 123–144.

[21] Council on Ethical and Judicial Affairs, American Medical Association, "Medical Futility in End-of-Life Care", *JAMA* 281 (1999), 938.

[22] For example, G. G. Griener, "The Physician's Authority to Withhold Futile Treatment", *The Journal of Medicine and Philosophy* 20 (1995), 209–210; K. Shiner, "Medical Futility: A Futile Concept?", *Washington & Lee Law Review* 53 (1996), 827.

[23] The Hastings Center, *Guidelines on the Termination of Life-Sustaining Treatment and the Care of the Dying* (Briarcliff Manor, NY: The Hastings Center, 1990), p. 32.

[24] R. Tong, "Towards a Just, Courageous, and Honest Resolution of the Futility Debate", *The Journal of Medicine and Philosophy* 20 (1995), 174, italics and peculiar punctuation in original.

[25] E. H. Morreim, "Profoundly Diminished Life: The Casualties of Coercion", *Hastings Center Report* 24 (1994), 34–35.

Why do physicians and patients usually focus their attention on the physiological effects of medical treatment? It is because achieving these ends normally contributes to the further ends of the patient. Thus, by means of administering antibiotics, a physician can often cure the patient's infection and in turn enable the patient to return to work, or CPR may be effective in restoring pulmonary and respiratory function and thereby restore the patient to an active worthwhile life. Lance Stell has pointed out the way in which the ends of diagnosis and therapy are "nested" in further ends from which they derive their value.[26] The lesson we should learn from this is that physiological futility is important only when it is associated with qualitative futility. What makes any demand for treatment rational is its promise of beneficial results, whatever its physiological effects or lack of them.

(2) When treatment would not benefit the patient in any way, then it is completely pointless and there is no reason why the physician ought to provide it or for the patient to demand it. Therefore, the demand for nonbeneficial treatment is irrational and the physician may disregard it. For example,

> Having suffered cardiorespiratory arrest, Sonya Causey was transferred to St. Francis Medical Center (SFMC) from a nursing home. She was comatose, quadriplegic and in end-stage renal failure. Her treating physician, Dr. Herschel R. Harter, believed that continuing dialysis would have no benefit. Although Dr. Harter agreed that with dialysis and a ventilator Mrs. Causey could live for another two years, he believed that she would have only a slight (1% to 5%) chance of regaining consciousness.[27]

Thus, the physician believed and Saint Francis Medical Center argued that although the treatment demanded by her husband would not be physiologically futile, they ought to be at liberty to refuse to provide it because it would be of no benefit. Presumably, this was because nothing could be of benefit to an unconscious patient. This assumption is made explicit in the conclusion of Helga Wanglie's attending physician that her respirator was nonbeneficial primarily because it would not "enable this unconscious and permanently respirator-dependent woman to experience the benefit of the life afforded by respirator support."[28]

One must recognize that not everyone accepts this assumption. As Marcia Angell observed:

> In the case of Helga Wanglie, the institution saw the respirator as "non-beneficial" because it would not restore her to consciousness. In the family's view, however, merely maintaining life was a worthy goal, and the respirator was not only effective toward that end, but essential. Public opinion polls indicate that most people would not want their lives maintained in a persistent vegetative state. Many consider life in this state to be an indignity, and care givers often find caring for such patients demoralizing. It is important,

---

[26] L. K. Stell, "Stopping Treatment on Grounds of Futility: A Role for Institutional Policy", *Saint Louis University Public Law Review* 11 (1992), 490–491.
[27] *Causey* at 1073.
[28] S. H. Miles, *Op. Cit.*, p. 513.

however, to acknowledge that not everyone agrees with this view and it is a highly personal issue.[29]

Although I believe that nothing could possibly benefit a permanently unconscious patient, I do not insist that the contrary opinion is irrational. There are some plausible reasons for believing that all human life is sacred irrespective of its experiential quality or that respect for the biological life of each and every person is morally required. Fortunately, the issue of irrationality may not matter. The crux of a more plausible argument for the physician's liberty to refuse treatment is not that the patient's demand is irrational, but that the requested treatment would be of no benefit.

Still, how often would a demanded treatment offer no benefit of any kind? Only, I suspect, when the patient is irreversibly comatose. But physicians judged aggressive medical treatment nonbeneficial in the cases of Baby K and J (a minor) although neither was in a persistent vegetative state. And a patient's request for a fifth or sixth series of chemotherapy treatments when the most optimistic prognosis is dismal might well be thought to be futile even though it might satisfy the patient's desire to remain alive for a few more weeks. The patient's request, or the request of a loving family member acquainted with the patient's values, is presumably some evidence that the demanded treatment would be of some value to the patient. Therefore, the judgment that the demanded treatment would not benefit the patient in any way is usually inapplicable to cases of alleged futility.

More generally relevant is the judgment that the demanded treatment would be nonbeneficial because at best the very slight benefits it promises are at least equaled by attendant harms. Although the demanded treatment may not be completely futile in that it would provide no benefit at all, it is futile on balance because it would provide no net benefit when the benefits that would result are weighed against the accompanying harmful consequences. Therefore, there is no sufficient reason why the physician should provide the requested medical treatment. And in cases where the demanded treatment would on balance do no good for the patient, there is no good reason why the law ought to require the physician to provide such treatment. Therefore, the law ought to leave the physician at liberty to refuse to provide it because of the moral principle that the law ought not to limit individual liberty without a good reason. There is a moral presumption and ought to be a legal presumption in favor of individual liberty because of the great value of individual liberty for the individual and for society. Since a legal liberty is simply the absence of any contrary legal duty, this implies that the physician should have a legal liberty to refrain from providing demanded medical treatment that is on balance futile.

But if benefit to the patient would be a reason to provide that treatment and normally to impose a duty to provide it upon the physician, why would not the benefit to others of a demanded treatment also justify requiring the physician to provide the demanded treatment? Well, a principle in medical ethics has traditionally been that the physician's moral duty is solely to the patient so that the interests of third parties should not intrude into the physician's clinical decision-making.[30] Although I am not convinced that it is

---

[29] M. Angell, "The Case of Helga Wanglie: A New Kind of 'Right to Die' Case", *The New England Journal of Medicine* 325 (1991), 512.
[30] Griener, *Op. Cit.*, p. 219.

morally impermissible for the physician to consider the benefits or harms to others of any contemplated treatment of the patient, I do believe that the physician's primary duty is to the patient. By accepting someone as a patient, the physician acquires a special duty to that person, a duty she does not owe to any third party. And because there may be a conflict of interests, or at least differences in value judgments, between a patient and other members of her family, the physician ought not to take demands of the family as seriously at requests of the patient herself. Therefore, although it may be permissible for a physician to consider benefits to persons other than the patient and in some cases to exercise her discretion to provide treatment that is futile, benefits to others are not a sufficient justification for the law to restrict her liberty to refuse to provide futile medical treatment demanded by a patient or her family.

(3) Another reason for a legal liberty of the physician to refuse to provide treatment demanded by the patient or her family is that the requested treatment would be abusive or inhumane. This argument was used by Scottish Rite Hospital in the case of Jane Doe after her mother consented to a DNR order, but her father did not.

> The hospital filed a declaratory judgment action seeking guidance as to which of the parent's wishes it should follow. Although the hospital did not allege child abuse, or seek to cut off the parents' custodial rights, it alleged that continued aggressive treatment of the child constituted medical abuse.[31]

Although this is an excellent reason for the law to permit the physician to refuse demanded medical treatment, strictly speaking it is not an appeal to medical futility. It is one thing to appeal to the fact that the requested treatment would be futile because nonbeneficial; it is quite another to argue that it is abusive or harmful.[32] The former suggests that there is no reason why the physician ought to provide the treatment; the latter shows that there is a reason why the physician ought not to provide it.

Nevertheless, the fact that the demanded treatment would be abusive or inhumane is not always irrelevant to a futility argument. The harms imposed upon the patient by such treatment might offset the benefits the requested treatment would provide so that it would provide no net benefit. In such cases, one could argue that the physician ought to have a legal liberty to refuse to treat as demanded because the treatment would be futile on balance.

(4) The physician ought to have a legal liberty to refuse to provide demanded medical treatment that is futile because futile treatment is a useless waste of medical resources. But why not waste medical resources? For one thing, medical resources are scarce. This implies that wasted medical resources will not be available to use for patients who could be benefited thereby.[33] Hence, patients thus denied treatment would be harmed. Also the medical resources available in our society to physicians and hospitals are subsidized to a considerable extent from public funds. These include, among other factors, payments under Medicare or Medicaid, governmental support for medical education and public funding for medical research. Therefore, it would be a violation of public trust

[31] *In re Jane Doe* at 4.
[32] R. Macklin, *Enemies of Patients* (New York & Oxford: Oxford University Press, 1993), p. 168.
[33] See *Re J (a minor)* [1992] 4 All ER 614 CA at 625.

for the physician[34] to waste medical resources that are paid for in part from the public purse.[35]

This argument is often misrepresented. It does not require any expansion of the concept of medical futility. It does not assume that some medical treatments are futile because they would be wasted. It argues in the opposite direction that some medical treatments would be wasted because they are futile. Again, it need not appeal to any principle of justice and argue that to provide futile medical treatment would be an unfair or inequitable allocation of medical resources. On the one hand, it appeals to the principle that physicians ought not to be required by law to harm those patients who would be denied treatment because they have wasted scarce medical resources. And on the other hand, it presupposes the principle that physicians owe a duty of trust to the public to the extent that the medical resources they use are subsidized by the public.

Ruth Macklin challenges the assumption that providing allegedly futile treatment to one patient would in fact deny treatment to another patient.

> It is probably safe to assume that Baby L was not occupying a bed or using life-support technology that was being denied another infant with "a reasonable chance of survival." The only other interpretation of this oft-repeated claim that resources are unavailable to others is that money saved by not treating Baby L would be used to save Baby M or N or O, who "have a reasonable chance of survival." But is there any evidence for this supposition? Baby L's medical expenses were covered by third-party payment. Are there any grounds for believing that the insurance company denied reimbursement to another child (or adult) because the cost of Baby L's treatment absorbed a large share of the insurance pool? Insurance doesn't work that way.[36]

Unfortunately, this optimistic rebuttal overlooks the long-term consequences and the indirect costs of wasting scarce medical resources. Although there may not have been another patient who at that time needed the hospital resources provided to Baby L, her use of the hospital's medical resources would be only in part reimbursed by payments from the insurance company. There are significant indirect costs that must also be taken into account.[37] And if the insurance company does not directly deny treatment to others in order to limit its expenditures, then it must increase its premiums thus increasing the number of persons who are denied medical treatment because they cannot afford medical insurance. Of course, the government might in turn extend Medicaid and Medicare to provide coverage to all members of our society. But this is probably politically out of the question, and were it to happen, taxpayers would be compelled to absorb the costs of wasted medical resources.

---

[34] But see R. M. Veatch and C. M. Spicer, "Medically Futile Care: The Role of the Physician in Setting Limits", *American Journal of Law & Medicine* 18 (1992), 28–29. They argue that it is the physician's role to benefit the patient, not to protect the interests of society. Hence, it ought to be society, not the physician, that decides when to limit medical treatment. However, the physician's duty to benefit her patient does not apply to treatment that would provide no net benefit, and when society has not ruled out wasteful treatment, the physician must accept her responsibility in this matter.

[35] Compare Miles, *Op. Cit.*, p. 514; Morreim, *Op. Cit.*, p. 35.

[36] Macklin, *Op. Cit.*, pp. 169–170.

[37] See Strasser, *Op. Cit.*, p. 545.

(5) Physicians and other medical practitioners ought to have the legal liberty to refuse to provide futile medical treatment demanded by their patients in order to preserve their professional integrity. Although the multiple physician–patient relationships and commercialization of modern medicine often weaken any appeal to professional integrity,[38] this remains a valuable ideal well-worth preserving. The professions have traditionally, and I believe rightly, been distinguished from trades and businesses by two defining features. The purpose of any profession is to serve the best interests of its clients rather than to make a profit, and in order to do so some branch of advanced specialized knowledge is required. For example, although someone without a thorough knowledge of medical science and extensive clinical experience could pretend to practice medicine, only one with a high degree of medical expertise could in practice serve the patient's best medical interests. The United States Supreme Court has recognized this:

> Few professions require more careful preparation by one who seeks to enter it than that of medicine. It has to deal with all those subtle and mysterious influences upon which health and life depend .... The physician must be able to detect readily the presence of disease, and prescribe appropriate remedies for its removal. Everyone may have occasion to consult him, but comparatively few can judge of the qualifications of learning and skill which he possesses. Reliance must be placed upon the assurance given by his license, issued by an authority competent to judge in that respect, that he possesses the requisite qualifications.[39]

It follows that licenses to practice medicine should be issued by competent members of the medical profession and that each physician ought to have professional autonomy, the freedom to practice in the light of his or her best professional judgment, in order to serve the best interests of the patient.

Some such reasoning underlies the 1986 *Declaration on Physician Independence and Professional Freedom* of the World Medical Association.

> Physicians must have the professional freedom to care for their patients without interference. The exercise of the physician's professional judgment and discretion in making clinical and ethical decisions in the care and treatment of patients must be preserved and protected.

This principle was accepted by an English Court of Appeal, Civil Division, in a medical futility case. Lord Donaldson argued:

> The fundamental issue in this appeal is whether the court in the exercise of its inherent power to protect the interests of minors should ever require a medical practitioner or health authority acting by a medical practitioner to adopt a course of treatment which in the bona fide clinical judgment of the practitioner concerned is contraindicated as not being in the best interests of the patient. I have to say that I cannot at present conceive of any circumstances in which

[38] See S. J. Youngner, "Medical Futility and the Social Contract (Who are the Real Doctors on Howard Brody's Islands?)" *Seton Hall Law Review* 25 (1995), 1015–1026.
[39] *Dent v. West Virginia*, 129 U.S. 114 (1889) at 122–123.

> this would be other than an abuse of power as directly or indirectly requiring the practitioner to act contrary to the fundamental duty which he owes to his patient. This, subject to obtaining any necessary consent, is to treat the patient in accordance with his best clinical judgment ....[40]

Here, the ethical requirement of physician autonomy is used to limit the legal duty of the physician to the patient in such a way that he or she has the legal liberty to refuse to provide futile medical treatments.

Although I believe that the need for a knowledge of medical science and wide clinical experience shows that the law ought to allow a considerable degree of physician autonomy, it does not follow that the discretion of each individual physician ought to be unlimited. In upholding a statute regulating the revocation of one's license to practice medicine, the Supreme Court of North Carolina reasoned that:

> The provision of the statute in question here is reasonably related to the public health. We conclude that the legislature ... reasonably believed that a general risk of endangering the public is inherent in any practices which fail to conform to the standards of "acceptable and prevailing" medical practice in North Carolina.[41]

Accordingly, it would be justifiable for the legal liberty of the physician to refuse to provide futile treatment demanded by the patient or her family to be limited by the requirement that it not greatly deviate from the accepted practice of the profession in general.[42]

(6) Physicians ought to have the legal liberty to refuse to provide futile medical treatment demanded by the patient or her family in order to preserve their moral integrity. Physicians are not merely members of the medical profession; much more importantly, they are individual moral agents. As Jecker and Schneiderman argue:

> The provision of futile treatment is additionally objectionable when the act violates a physician's personal ethical convictions. In this case, a refusal to allow the physician to withhold or withdraw futile interventions does not take seriously the physician's own ethical autonomy and agency. It would be akin to mandating that physicians who oppose abortions perform them. In these cases, requiring the use of futile interventions wrongly signals that physicians are merely tools for enacting others' (patients') goals and do not possess, as individuals and as members of a profession, independent ethical standards and ends.[43]

Therefore, the law ought to leave physicians free to refuse on conscientious grounds to provide futile medical treatment.

But why should the law permit conscientious objections to limit the physician's duty to provide treatment demanded by a patient or her family? Well, for one thing, there is an important state interest in having citizens who take moral considerations seriously

---

[40] *Re J (a minor),* at 622.
[41] *In Re Guess*, 393 S.E.2nd 833 (N.C.1990) at 837.
[42] See *Matter of Dinnerstein*, Mass. App., 380 N.E.2nd 134 (1978) at 139.
[43] Jecker and Schneiderman, *Op. Cit.*, pp. 153–154.

and act accordingly. This is especially true of citizens, like medical practitioners, who occupy the most important roles in society. For another, any attempt to enforce a duty to provide treatment, a physician considers morally wrong would have harmful social consequences. It would transform the physician–patient relationship from a commitment to bring one's best knowledge and skill to help the patient meet important needs into an irrevocable indentured servitude to the demanding patient.[44] And it would discourage the most conscientious and dedicated persons from entering the medical profession or remaining in it.[45]

But are there ever any genuine moral reasons not to provide futile medical treatment? There certainly seem to be several.[46] A patient might demand medical care inconsistent with the proper goals of medicine, for example, ineffective treatments that have harmful side effects. The interests of family members sometimes conflict with those of a patient so that the family might request medical treatment that would not be in the best interests of an incompetent patient for whom they are acting. And it is not unreasonable to suppose that it is unethical to waste scarce medical resources by using them for futile medical treatments. Although not everyone will share these moral convictions, presumably the law ought to respect the moral integrity of a sincere moral agent even when his or her conscience may be mistaken.[47]

There are, then, at least six reasons one might well give for the conclusion that physicians and other medical practitioners ought to have the legal liberty to refuse to provide futile medical treatment demanded by patients or their families. No doubt others have been advanced in the legal literature, but these strike me as by far the most plausible. I have often reformulated the arguments to make their logical structure explicit and their premises more reasonable. Of these, I find the arguments that the physician ought to have this legal liberty because the demanded treatment would provide no net benefit, it would be a waste of scarce medical resources, in order to preserve the physician's professional integrity and the physician's moral integrity the most cogent. Together they make a very strong case for the legal liberty of physicians and other medical practitioners to refuse to provide futile medical treatment demanded by their patients.

10.3. Reasons Against a Legal Liberty

There are, as one would expect, at least as many plausible arguments for the opposite conclusion. (1) The physician ought not to have a legal liberty to refuse to provide futile medical treatment demanded by the patient or her family because any such denial would constitute a violation of the patient's right not to be abandoned.

---

[44] Morreim, *Op. Cit.*, p. 37.

[45] E. W. Clayton, "Commentary: What Is Really at Stake in Baby K? A Response to Ellen Flannery", *Journal of Law, Medicine & Ethics* 23 (1995), 14.

[46] Compare Shiner, *Op. Cit.*, pp. 834–835.

[47] But see R. M. Veatch and C. M. Spicer, *Op. Cit.*, pp. 26–28. They argue that the physician's moral autonomy is limited by her duty of fidelity to her patient. Although the relationship of physician to patient does impose a special duty of fidelity, I believe that this duty is limited to the provision of care that has at least some net benefit to the patient.

> 5. The patient has the right to continuity of health care .... The physician may not discontinue treatment of a patient as long as further treatment is medically indicated, without giving the patient reasonable assistance and sufficient opportunity to make alternative arrangements for care.[48]

Presumably, the patient does have such a moral right grounded either on her contractual understanding with her physician or on her reliance upon his continuing care. But this right is probably inapplicable to most cases of medical futility. In the case of Baby K, for example,

> The physicians explained to Ms. H that most anencephalic infants die within a few days of birth due to breathing difficulties and other complications. Because aggressive treatment would serve no therapeutic or palliative purpose, they recommended that Baby K only be provided with supportive care in the form of nutrition, hydration, and warmth.[49]

Refusing to provide futile medical treatment does not normally constitute abandoning the patient because the physician continues to provide other forms of medical care.

Still, one might argue that to refuse demanded medical treatment would constitute neglect of the patient. In the opinion of the American Medical Association:

> 8.11 Neglect of Patient. Physicians are free to choose whom they will serve .... Once having undertaken a case, the physician should not neglect the patient, nor withdraw from the case without giving notice to the patient, the relatives, or responsible friends sufficiently long in advance of withdrawal to permit another medical attendant to be secured.[50]

I very much doubt, however, that to refuse to provide genuinely futile medical treatment would be to neglect the patient. To neglect a patient is to fail to provide medically indicated treatment, treatment needed by the patient for medical reasons. It is morally wrong to neglect a patient because to fail to provide needed care would be to harm the patient, or at least to risk harming her. But providing futile medical case would not benefit the patient, and failing to provide it would not harm her either. Hence, this argument rests upon mistaken premises because refusing to provide demanded medical treatment that really is futile is neither to abandon nor to neglect one's patient.

(2) The physician ought not to have a legal liberty to refuse to provide futile medical treatment demanded by the patient or her family acting as her surrogate because to refuse her request would be to fail to respect her autonomy. As Keith Shiner reminds us:

> Courts and legislatures have placed the decisionmaking authority with patients primarily out of respect for autonomy or self-determination. Thus, an argument that physicians have the authority to declare certain requested treatments futile must overcome patients' autonomy rights as courts have developed them over the last twenty years and demonstrate why physicians have a greater authority to

---

[48] American Medical Association, *Fundamental Elements of the Patient–Physician Relationship*, 1990, updated 1993.

[49] Baby "K" at 592.

[50] American Medical Association, *Current Opinions of the Council on Ethical and Judicial Affairs*, 1994.

> decide the use of care at the extremes of life. A competent individual generally has the right to control what is done with his or her body. Courts base this right upon various grounds, but they generally rely either on the common law of informed consent or on a constitutional privacy right.[51]

What Shiner and other commentators fail to notice, however, is that these two grounds of patient autonomy concern two very different aspects of autonomy. The common law of informed consent was originally an application of the right not to be assaulted based upon the legal and moral right of personal security.[52] Hence, it concerns one's autonomy as a patient (in the grammatical sense of one who is acted upon by another). On the other hand, the constitutional right to privacy is the right that others, especially the state, not intrude upon or interfere with one's liberty of making and acting upon private decisions.[53] Hence, it concerns one's autonomy as an agent (a moral agent who has the capacity to choose and act rationally).

Let us consider first one's autonomy as a patient, as someone acted upon by the physician. Because any medical treatment imposes some risk of physical or psychological harm, one has a right not to be treated by one's physician unless one waives one's legal and moral right to personal security by giving one's consent. And this requirement of consent to medical treatment implies that one's refusal to consent to any recommended or threatened medical treatment imposes an even stronger duty of the physician to refrain from such treatment. It does not, however, imply any legal right to be provided with whatever treatment one requests.

> Legally, the doctrine of patient autonomy and the right to die are predicated on a patient's liberty interest in rejecting unwanted medical treatment and exist independently of any right of access to treatment. This principle was expressed as early as 1914 by Judge Cardozo in his assertion that "every human being of adult years and sound mind has a right to determine what shall be done with his own body ...." As Tomlinson and Brody and Veatch and Spicer noted, it does not follow that the existence of a right to reject unwanted treatment implies a similar right to receive desired treatment.[54]

Similarly, one's moral right to personal security including bodily integrity implies that the physician has a moral duty not to treat one without one's consent but does not imply any duty to provide whatever treatment one requests.

There might, of course, be some other moral right that would imply a duty to provide demanded medical treatment. For example, Article 12 of the 1966 *International Covenant on Economic, Social and Cultural Rights* recognizes "the right of everyone to the enjoyment of the highest attainable standard of physical and mental health." Attempts to identify the grounds of this or any comparable human right to health care have not, however, been successful.[55] Moreover, it is hard to see how any such human right would imply any duty

---

[51] Keith Shiner, *Op. Cit.*, pp. 812–813; see also, Marcia Angell, *Op. Cit.*, p. 511.

[52] See *Schloendorff v. Society of N. Y. Hosp.*, 105 N.E. 92, 93 (N.Y. 1914).

[53] See *Eisenstadt v. Baird*, 405 U.S. 438 (1972) at 453 and *In Re Quinlan*, 355 A. 2nd 647 (1976) at 663.

[54] D. H. Johnson, "Helga Wanglie Revisited: Medical Futility and the Limits of Autonomy", *Cambridge Quarterly of Healthcare Ethics* 2 (1993), 162. See also, Shiner, *Op. Cit.*, pp. 837–838.

[55] C. Wellman, *The Proliferation of Rights* (Boulder CO: Westview Press, 1999), pp. 155–159.

of the physician to provide futile medical treatment because futile care is by definition incapable of advancing the patient's physical or mental health in any way.

A more relevant right would seem to be the right to emergency medical care. The federal EMTALA does confer a positive right to be provided with medical care upon any patient diagnosed as presenting an emergency condition. On this basis, the United States Court of Appeals, Fourth Circuit, required a hospital to provide Baby K with aggressive medical treatment considered futile by its physicians.[56] One might well argue that there is a comparable moral right to be provided with medical treatment grounded on one's fundamental moral right to rescue. However, any such right would not seem to imply any right to be provided with futile medical treatment because completely futile treatment would be incapable of rescuing the patient from any danger of medical harm. Therefore, there does not seem to be any argument against a legal liberty of the physician or other medical practitioners to refuse to provide demanded medical treatment that is futile from the patient's autonomy as a patient.[57]

There remains the patient's autonomy as an agent, as a person capable of rational decision and action. Every patient has a constitutional right to privacy, a right that others, especially the state, not interfere with her acting on private decisions, including such decisions as to whether to seek and have an abortion or to refuse medical treatment. Similarly, I am willing to grant that she has a moral right to privacy, a right that others not prevent or hinder her from acting on private choices, including her choice of a physician and whether to consent to his medical recommendations or to request some alternative treatment. But a physician's refusal to provide medical treatment demanded by her does not prevent or hinder her from acting in any of these ways. Therefore, her autonomy as an agent, her right to act as she chooses, does not imply any duty of others to act as she desires. Indeed, their moral right as autonomous moral agents precludes this. Nevertheless, the patient's moral autonomy as an agent is not irrelevant to the futility debate. It does imply that the physician has a duty not to hinder or prevent his patient from seeking, and if available obtaining, treatment he considers futile. Therefore, any legal liberty of the physician to refuse to provide futile medical treatment demanded by a patient should be limited by the physician's legal duties to inform his patient of his refusal and to co-operate in enabling her to find another physician who would be willing to treat her as she requests.[58]

(3) The physician ought not to have a legal liberty to refuse to provide futile medical treatment demanded by the patient or her family because his refusal would often violate the patient's right to life. Arguably, this was true in the cases of Helga Wanglie and Jane Doe, and the reasoning of the United States District Court made this argument explicit in the case of Baby K.

> "When parents do not agree on the issue of termination of life support . . . this Court must yield to the presumption in favor of life." . . . This presumption arises from the explicit guarantees of a right to life in the United States Constitution,

---

[56] *In the Matter of Baby "K"*, at 594 and 598.

[57] Compare *Barber v. Superior Court*, 195 Cal. Rptr. 484 (Ct. App. 1983) at 493.

[58] Compare P. G. Peters, Jr., "When Physicians Balk at Futile Care: Implications of the Disability Rights Laws", *Northwestern University Law Review* 91 (1997) 798, at 842.

> Amendments V and XIV, and the Virginia Constitution, Article 1, Sections 1 and 11.[59]

Clearly, any justification for a physician's legal liberty to refuse to provide futile medical care demanded by the patient must overcome this legal presumption in favor of life.[60] There certainly seems to be a comparable moral presumption grounded on the human right to life.

As I have explained elsewhere, the human right to life is a rights-package, a set of fundamental moral rights concerning the life of each human being. These include at least the moral claim-right not to be killed by another, the moral claim-right that others not endanger one's life, the moral liberty-right to defend one's life with all necessary force, the moral liberty-right to preserve one's life by any necessary means, and perhaps the moral claim-right to be rescued from the danger of death.[61] Were a physician to refuse to continue life support treatment, one could plausibly argue that this violates the patient's moral claim-right not to be killed by another, or at least her moral claim-right not to have her life endangered. The physicians attending Jane Doe advanced one reply:

> The doctors agreed that she lacked the ability for any cognitive function or interactive activity, and did not have any reasonable hope for recovery. They also agreed there was no known medical treatment that could improve her condition or halt the neurological deterioration. It was apparent that the life support system was prolonging her death, rather than her life.[62]

But this reply is formulated in misleading language. The life support system was not prolonging her death, it was prolonging her dying. And dying is a process that takes place during the last period in one's life, at least in the ordinary sense of the word "life." To be sure, the hospital did allege that "continued aggressive treatment of the child constituted medical abuse."[63] But any such appeal to the harm imposed by treatment is quite different from justifying the liberty to deny futile, nonbeneficial treatment that seems to violate the patient's right to life.

A more adequate argument that I believe does overcome the presumption in favor of life was suggested in the case of Helga Wanglie:

> In October 1990, a new attending physician consulted with specialists and confirmed the permanence of the patient's cerebral and pulmonary conditions. He concluded that she was at the end of her life and that the respirator was "non-beneficial," in that it could not heal her lungs, palliate her suffering, or enable this unconscious and permanently respirator-dependent woman to experience the benefit of the life afforded by respirator support.[64]

---

[59] *In the Matter of Baby K* at 1030.
[60] Shiner, *Op. Cit.*, p. 811.
[61] Carl Wellman, *An Approach To Rights*, pp. 245–248.
[62] *In re Jane DOE* at 6.
[63] *Ibid.*, at 4.
[64] Miles, *Op. Cit.*, p. 513.

One way of interpreting the force of this suggestion is to argue that although a patient in an irreversible coma may be biologically alive, she is not alive in the sense relevant to the legal or moral right to life. In the case of Nancy Cruzan, Justice Stevens argued:

> Nancy Cruzan is obviously "*alive*" in a physiological sense. But for patients like Nancy Cruzan, who have no consciousness and no chance of recovery, there is a serious question as to whether the mere persistence of their bodies is "*life*" as that word is commonly understood, or as it is used in both the Constitution and the Declaration of Independence .... Life, particularly human life, is not commonly thought of as a merely physiological condition or function. Its sanctity is often thought to derive from the impossibility of any such reduction. When people speak of life, they often mean to describe the experiences that comprise a person's history, as when it is said that somebody "led a good life."[65]

I find this argument convincing and believe that what qualifies one for the moral right to life is that one has a life, a conscious biography, not merely biological existence as a human organism.[66] If Justice Stevens and I are correct, then at least for those patients who are permanently unconscious, there seems to be no legal and moral presumption in favor of life that the denial of futile medical treatment needs to overcome.

(4) The physician ought not to have a legal liberty to refuse to provide futile medical treatment demanded by the patient or her family because his refusal would often violate the patient's right to nondiscrimination. The United States District Court held that this was so in the case of Baby K:

> The Hospital asks this court for authorization to deny the benefits of ventilator services to Baby K by reason of her anencephaly. The Hospital's claim is that it is "futile" to keep alive an anencephalic baby, even though the mother has requested such treatment. But the plain language of the ADA does not permit the denial of ventilator services that would keep alive an anencephalic baby when those life-saving services would otherwise be provided to a baby without disabilities at the parent's request. The Hospital's reasoning would lead to the denial of medical services to anencephalic babies as a class of disabled individuals. Such discrimination against a vulnerable population class is exactly what the American with Disabilities Act was enacted to prohibit.[67]

Because the federal antidiscrimination laws define "disability" very broadly, this judicial reasoning would be applicable to a wide variety of medical futility cases.[68]

The most obvious justification for statutes that prohibit discrimination of any kind is that to discriminate against any individual is to violate her fundamental moral right to equitable treatment. This is the moral claim-right against second parties that they not treat one worse than they treat other comparably situated individuals unless there is some reason that this different treatment is not unjust.[69] Now, there are two plausible ways to rebut

---

[65] *Cruzan v. Director, Mo. Health Dept.*, 497 US 261 (1990) at 345–346, italics in original.

[66] Carl Wellman, *The Proliferation of Rights*, pp. 98–99.

[67] *Matter of Baby K* at 1029.

[68] Peters, *Op. Cit.*, p. 807.

[69] Compare C. Wellman, *Welfare Rights*, pp. 139–147.

the charge that the physician's refusal to provide futile medical treatment is inequitable or discriminatory in any morally relevant ways. First, one might argue that the patient refused some sort of treatment is not similarly situated with those patients who receive this kind of medical treatment because she has no moral claim to such treatment while those who receive it do have this moral claim. To rebut this argument, one would have to specify some prior and independent ground of the patient's moral claim to receive futile medical treatment. Second, and more directly, one can argue that to refuse to provide futile medical treatment to a patient is not to treat that patient worse than one treats other patients because futile medical treatment is by definition nonbeneficial. Hence, to deny genuinely futile medical treatment to a patient is not to discriminate against her by treating her worse than one treats other patients because one is not thereby denying her benefits one provides to others who are similarly situated. To my mind, this reasoning is cogent and implies that the courts ought to interpret the federal antidiscrimination laws so that they are compatible with the physician's legal liberty to refuse to provide medical treatment that really is futile.[70] Where the District Court went astray was to assume that the hospital was "asking for authorization to deny the benefits of ventilator services to Baby K." Because the hospital regarded those services nonbeneficial, it was not asking for authorization to deny benefits to Baby K.

(5) The physician ought not to have a legal liberty to refuse to provide futile medical treatment demanded by the patient or her family because no objective and concrete definition of futility is possible. The Council on Ethical and Judicial Affairs of the American Medical Association reported "Futility is an essentially subjective but realistically indispensable judgment. A fully objective and concrete definition of futility is unattainable."[71] Presumably, the concept of futility is needed in order for the physician to advise his patient regarding the available treatment options and probably for the patient to make a morally justified decision whether to consent to having treatment her physician judges futile withheld or withdrawn. But for legal purposes, one needs a definition of futility that is concrete enough so that it can yield a definite decision when applied to the facts of individual cases and objective enough so that disagreements between physician and their patients can be resolved satisfactorily. If no such definition is possible, it would be useless or even counter-productive to introduce any liberty defined in terms of futility into our legal system.

This was the conclusion that the Second Circuit Court of Appeals of Louisiana reached in the light of a number of previous medical futility cases.

> Futility is a subjective and nebulous concept which, except in the strictest physiological sense, incorporates value judgments. Obviously, in this case, subjective personal values of the benefit of prolonging life with only a slight possibility of improvement, dictated SFMC's and Dr. Harter's decision. To focus on a definition of "futility" is confusing and generates polemical discussions. We turn instead to an approach emphasizing the standard of medical care.[72]

---

[70] See also, E. H. Morreim, "Futilitarianism, Exoticare, and Coerced Altruism: The ADA Meets Its Limits", *Seton Hall Law Review* 25 (1995), especially pp. 897–899.

[71] "Medical Futility in End-of Life Care", *JAMA* 281 (1999), 938. See also, B. A. Brody and A. Halevy, "Is Futility a Futile Concept?", *Journal of Medicine and Philosophy* 20 (1995), 123–144.

[72] *Causey* at 1075.

Since polemical arguments are useless in judicial reasoning capable of settling legal disputes, the law ought not to define legal rights or duties in terms of such a subjective and nebulous concept as futility. To add the concept of medical futility to the law would introduce a high degree of legal uncertainty that would be unfair to both physicians and patients because prior to a judicial decision the former could not know whether they were legally at liberty to refuse medical treatment in any given case and the latter could never know whether they had a legal claim to receive treatment they desire but their physicians judge futile.[73]

In what sense of "subjective" might the concept of medical futility be subjective? It might be relative to each individual subject rather than applying universally to all subjects with the same medical diagnosis and prognosis. Some of those who believe that this is so propose that medical futility be defined in terms of the patient's values or goals.[74] But because different patients have very different goals in life and even regarding their deaths, whether some sort of medical treatment is futile will vary from one patient to another.

Now because to say that something is futile is to say that it is incapable of achieving its purpose, futility is relative to some assumed goal. But medical treatments do not have goals; only patients and their physicians have goals. However, I have argued that futile medical treatment ought to be defined as treatment that is nonbeneficial in the sense that it has no net benefit for the patient. If one accepts this definition, then the futility of any available medical treatment is defined in terms of a single goal, benefit to the patient. Nevertheless, critics suggest that this universality obscures an underlying relativity because whether something is beneficial for a patient depends upon the patient's goals. This suggestion presupposes something like a preference theory of value, a theory I reject.[75] Desiring or wanting something does not necessarily ensure its value as one far too often discovers by achieving some goal and finding it worthless or even undesirable. Examples might be seeking and gaining revenge only to find that it has destroyed a valuable friendship or obtaining a position of corporate leadership that has unsuspected responsibilities and disadvantages. I do not wish to deny that a person's goals have some relevance to her well-being. I deny only that they entirely determine it by the very nature of what is of value to a person. Granted that what benefits a person will be to some extent relative to each individual patient, there will in every case be one correct judgment about whether any demanded treatment is or is not futile for that patient given that patient's diagnosis, prognosis, and values.

But will patient, physician, and judge be able to recognize the truth of that judgment? The concept of medical futility may still be subjective in the sense that reasonable subjects cannot overcome their disagreement about whether or not some demanded medical treatment is in fact futile. The medical futility cases that have gone to litigation show that physicians and patients have often been unable to overcome their disagreement in the most serious of circumstances. And the way that the judicial reasoning in most of these cases has evaded the issue of futility suggests that judges are in no better position to reach convincing conclusions about medical futility. It seems as though disputes about medical

---

[73] Compare M. B. Kapp, "Commentary: Anxieties as a Legal Impediment to the Doctor–Proxy Relationship", *Journal of Law, Medicine & Ethics* 27 (1999), 69–73.

[74] Strasser, *Op. Cit.*, p. 527.

[75] See also L. K. Stell, *Op. Cit.*, pp. 488–489.

futility are subjective in the sense that they cannot be resolved by rational discussion because moral reasoning must always begin with assuming that certain kinds of things are or are not valuable.[76] I have argued elsewhere that it is a mistake to reduce all ethical justification to deductive reasoning that takes some universal judgment of value or principle of obligation for granted.[77] Because there are other forms of ethical argument, it is possible in principle to resolve all moral disagreements. In practice, of course, agreement will not always be achieved. However, the law considers benefits and harms to patients in other areas of medical law, for example, when deciding whether a patient has grounds to sue for medical malpractice and in determining the award for damages when medical malpractice has been found, so that there is no conclusive reason to believe that it could not deal with medical futility in some similar manner.[78]

(6) The physician ought not to have a legal liberty to refuse to provide futile medical treatment demanded by the patient or her family because any such legal liberty would be subject to intolerable abuse. Because medical futility cannot be defined with precision, it would be impossible to define the limits of any legal liberty to refuse futile medical treatment clearly. As a consequence, over-zealous physicians will often deny treatment to patients when this is morally, and perhaps legally, unjustified. Medical futility becomes a moral or legal issue only when physicians are unable to convince their patients that some requested treatment would be nonbeneficial for them. Such disputes rest upon fundamental value judgments about the quality of life that moral reasoning seems incapable of resolving.[79] Glen G. Griener describes some of the abuses this invites.

> Placing authority in the hands of the patient's physician has some clearly recognized dangers. Physicians may make decisions on the basis of their own values where these do not coincide with the patient's. The practice may lead to insidious discrimination against those with physical or mental disabilities, or even against racial minorities or the economically disadvantaged. It may also mean that whether patients receive treatment depends upon who their physician happens to be, rather than on their medical condition.[80]

There is also the danger that physicians might claim that a treatment is nonbeneficial when they really believe that it is simply not worth the cost.[81] Marcia Angel asserts:

> It is dismaying, of course, that resources are spent sustaining the lives of patients who will never be sentient, but we as a society would be on the slipperiest of slopes if we permitted ourselves to withdraw life support from a patient simply because it would save money.[82]

That a legal liberty of the physician to refuse to provide medical treatment demanded by his patient might be abused in any or all of these ways is certainly possible.

---

[76] See Morreim, *Op. Cit.*, p. 34.
[77] C. Wellman, *Challenge and Response* (Carbondale IL: Southern Illinois University Press, 1971), pp. 51–83.
[78] See also Peters, *Op. Cit.*, pp. 832–833.
[79] Morrein, *Op. Cit.*, p. 34.
[80] G. G. Griener, "The Physician's Authority to Withhold Futile Treatment," *The Journal of Medicine and Philosophy* 20 (1995), 212.
[81] Peters, *Op. Cit.*, p. 856.
[82] Angell, *Op. Cit.*, p. 512.

Although these potential abuses cannot be entirely prevented, I believe that they could be reduced to a tolerable level by several practical proposals. Hospitals and other health care institutions should require a physician to consult its ethics committee before refusing to provide medical treatment requested by his patient. And the patient or her family should have the legal power to appeal a physician's refusal to provide demanded medical treatment to the courts. In such a case, the physician's personal judgment of medical futility should be subject to review in the light of the standard of good medical practice by the profession at large. These measures would reduce, although probably not entirely eliminate, the degree of arbitrary subjectivity concerning medical futility. Finally and of greatest importance, the patient should have the legal liberty to find a physician willing and able to provide the medical treatment she desires, and her physician should have a legal duty to co-operate with her in this regard.

(7) The physician ought not to have a legal liberty to refuse to provide futile medical treatment demanded by the patient or her family because there are better ways of solving the problems posed by the demand for futile treatment. These are that the emphasis upon patient autonomy in contemporary medical law and biomedical ethics is forcing physicians to provide medical treatments that are of no benefit to their patients or even burdensome to them because they merely prolong the process of dying, to waste medical resources that are needed by other patients who could benefit from treatment, and to violate their own sense of professional and moral integrity.[83]

One alternative would be to define the legal issue in terms of the standard of care in medical malpractice law rather than in terms of a liberty to refuse futile medical care. This was adopted in the opinion of the Louisiana Court of Appeal in the case of Sonya Causey.

> Futility is a subjective and nebulous concept which, except in the strictest physiological sense, incorporates value judgments .... To focus on a definition of "futility" is confusing and generates polemical discussions. We turn instead to an approach emphasizing the standard of medical care.[84]

Jerry Menikoff has advocated this approach more thoroughly than other jurists.

> This article will attempt to demonstrate that the current overwhelming attention being given to distinguishing futile from non-futile care is misplaced. Since physicians are subject to the rules of negligence law, as embodied in malpractice law, it is apparent that patients clearly *can* effectively demand elements of medical care that are required to meet the minimum malpractice standards. Thus, the interesting question relates not to futile care, but rather to a much broader category, namely care that is beyond those elements required by malpractice law, which we shall refer to as "extra-standard" care.[85]

His thesis is that even when the patient's demand is reasonable, her request for some kind of medical treatment should have no relevance to the physician's legal obligations. If the

[83] See, for example, Morreim, *Op. Cit.*, pp. 33–34; Brody and Halevy, *Op. Cit.*, pp. 123–125; and Shiner, *Op. Cit.*, p. 806.
[84] *Causey* at 1075.
[85] J. Menikoff, "Demanded Medical Care", *Arizona State Law Journal* 30 (1998), 1093, italics in original.

standard of medical care does not require the physician to offer it, then the law ought not to compel him to do so.[86] He points out that the practice of medicine is highly regulated and openly paternalistic because of the special vulnerability of patients. Physicians are and ought to be expected to use their independent medical judgment limited by social regulation for the good of their patients, not merely to inform patients who then have unlimited free choice among all possible medical treatments.[87] He also argues that if the patient's request for extra-standard care were to impose a duty to provide it, then the requirement of informed consent would expand to become impracticable.[88] Since there are strong moral reasons to preserve the legal requirement of informed consent, the physician should have no legal duty to provide extra-standard care merely because the patient requests it. In effect, what Menikoff proposes is that the physician's legal liberty to refuse to provide medical care demanded by his patient or her surrogate be defined as the liberty to refuse extra-standard care, not futile care.

But can one avoid confronting the issue of medical futility in its own terms? Menikoff's proposal seems to be confronted with a dilemma. Either the standard of medical care in medical malpractice law will incorporate the concept of medical futility or it will not. If it does, then the law will have recognized a legal liberty of physicians to refuse to provide futile medical care demanded by their patients so that this is not an alternative approach after all. If it does not, then the law will fail to apply to an essential aspect of medical practice. The Council Report of the American Medical Association asserts "Futility is an essentially subjective but realistically indispensable judgment."[89] Presumably, this means that physicians and other providers of medical care must distinguish between beneficial and futile medical care in order to inform their patients about the benefits and risks of possible medical treatments and to recommend what they judge to be the best available option. However, no fully objective and concrete definition of futility is attainable. For this reason, it does not recommend that medical law include any legal liberty to refuse futile medical care to patients who request it. Rather, "The Council, therefore, recommends that healthcare institutions, whether large or small, adopt a policy on medical futility and that policies on medical futility follow a fair process approach such as that presented above."[90] Here is a procedural rather than a substantive alternative to a legal liberty to refuse futile medical treatment.

Because they believe that attempts to define medical futility have failed, but the problem of making decisions about treatments that are of minimal benefit cannot be avoided, several physicians have proposed a solution in terms of a process of conflict resolution.

> The judgment that further treatment would be futile is not a conclusion—a signal that care should cease; instead, it should initiate the difficult task of discussing the situation with the patient. Thus, the most recent attempts to establish policy in this area have emphasized processes for discussing futility

---

[86] *Ibid.*, p. 1094.
[87] *Ibid.*, pp. 1114–1117.
[88] *Ibid.*, pp. 1121–1123.e
[89] AMA Council Report, "Medical Futility in End-of-Life Care", p. 938.
[90] *Ibid.*, p. 940.

> rather than the means of implementing decisions about futility. Talking to patients and their families should remain the focus of our efforts.[91]

Other authors have added that discussions between patients and their physicians need to be embedded within institutional procedures and policies.

> Conflict resolution processes that rely on consultative consensus-building, mediation, and counseling approaches should be encouraged as they appear well-suited to the complex decisions involved in health care relationships. Including such processes and resources explicitly in written hospital policies, especially at the point that medical futility is addressed, can remind providers that these avenues should be pursued first. Procedural ambiguity, however, can lead to confusion over authority and miscommunication about institutional decision-making. Hospitals should provide substantial opportunities for negotiation, mediation and consultation, but the policy should also specify a resolution mechanism for cases of persistent conflict.[92]

Because medical futility becomes a legal issue only when physicians and their patients disagree persistently about whether some requested treatment really is futile or beneficial, it would be better to institute procedures for resolving this disagreement rather than seeking a judicial decision that must inevitably be unsatisfactory to at least one of the parties to the dispute.

I heartily endorse the view that the first step in solving the problems posed by medical futility should be to institute processes of conflict resolution such as discussion, consultation, and counseling and that procedures for facilitating these processes should be written into the policies of hospitals and other healthcare institutions. But as Sandra H. Johnson et al. recognize, there will be some cases of persistent conflict that require an institutional resolution mechanism. Because it would be unjust to deny either party an opportunity to appeal such a resolution to the courts, this purely procedural alternative cannot be a complete solution to the problems of medical futility.

Susan B. Rubin does not believe that either of these alternatives goes far enough because they fail to recognize the most basic problems in any appeal to medical futility. The word "futility" is usually used without specifying the goals with respect to which a treatment is labeled futile, much less recognizing whose goals they are. It suggests that the issue is a purely medical issue within the competence of the physician and thereby obscures the need for public debate about the nature and scope of the authority that society grants to the medical profession. And any legal liberty of the physician to deny allegedly futile treatment would permit a unilateral decision making by the physician that would damage the trust essential to the therapeutic relationship, be an unjustified paternalistic treatment of the patient, and constitute an unwarranted exercise of power by the medical profession.[93]

Given the rapidly changing nature of modern medical practice, what is urgently needed is a clarification and perhaps a renegotiation of the social contract by which society gives physicians a license to practice and specifies the rules governing their practice.

---

[91] Helft et al., pp. 294–295.

[92] Johnson et al., p. 34. See also Tong, *Op. Cit.*, pp. 183–185.

[93] Susan B. Rubin, *When Doctors Say No* (Bloomington and Indianapolis: Indiana University Press, 1998), pp. 115–118.

> It has become commonplace to argue that medicine is an inherently moral enterprise. What has been missing in the futility debate is the recognition that medicine is also essentially, and unavoidably, a social practice. The practice of medicine exists within a particular social context, is informed inevitably and appropriately by the values of that context, and is constrained ultimately by the parameters set by that context. That is why the question of whether physicians should be empowered to make unilateral decisions on the basis of futility cannot be answered exclusively by the medical profession. Any question about the limits that should be set to the scope and practice of medicine must ultimately be answered not just by medicine but by society as well. Such a question must be asked and debated publicly, in the context of a critical and prior discussion about the nature of the relationship between the medical profession and society.[94]

Although this discussion has already begun in our courts and in scholarly publications, there has not yet emerged any widespread social consensus that has redefined the authority of practicing physicians with sufficient precision.

If demanded medical treatment is not to be denied on the grounds that it is futile, on what grounds may it be denied? One justification is that the requested treatment is irrelevant, that there is no causal connection between the treatment and the goal for which it is being considered. But for Rubin, this justification presupposes an overwhelming social consensus.

> Such strong social determination of a treatment's irrelevance may at first appear to be a fact known with a high degree of certainty. But as I use the term, relevance simply means that the evaluative component of the knowledge claim in question is supported by such overwhelming and widespread consensus, that it is rendered practically invisible. This leaves the impression that the claim is based exclusively on indisputable data. In fact, the refusal is justified not because it actually "is" absolutely irrelevant, but rather because we overwhelmingly agree that it is so. Our overwhelming agreement serves as an endorsement of the physician's refusal. The difference between claiming that a particular treatment is physiologically futile and claiming that it is irrelevant is nothing more than a difference in the degree of public consensus and endorsement.[95]

But on what ground might society intervene to deny a patient treatment that she considers of benefit to herself?

At this point, Rubin recognizes that any justified social consensus must take into account some more fundamental moral standard.

> With respect to overriding a patient's assessment of a treatment's worth, society has standing only on social justice grounds. In other words, society never has authority to determine a treatment's intrinsic worth for any given patient; the prerogative to make such an evaluative judgment is reserved for the patient

[94] *Ibid.*, p. 120.
[95] *Ibid.*, pp. 127–128.

> herself. On the other hand, if social consensus deems certain treatments to be insufficiently worth pursuing *with respect to other competing social goods*, rather than insufficiently worth it *per se*, a limit can justifiably be set to treatments individual patients can reasonably expect to receive. Only society, not the individual physician, or even the medical profession at large, can legitimately set limits on evaluative grounds to what individuals can expect to receive as parties to the social contract.[96]

Thus, ultimately the appeal must be to a social consensus concerning social justice and not medical futility.

I see two problems with this alternative. First, it will not work in practice. Granted the importance of rethinking, the social contract by which society licenses physicians to practice medicine, this must be an ongoing and never-ending renegotiation because of the constantly changing conditions of medical practice. It is unrealistic to imagine that in all or even most cases where a patient demands treatment that her physician judges of no net benefit, there will exist any social consensus to resolve their disagreement. Rubin herself recognizes this and suggests that in practice, genuine conversation between physicians and patients, the use of ethics committees, and a reference of intractable cases to the courts will be necessary.[97] Secondly, it misplaces the ground of any justifiable solution to the various problems posed by the current debate about medical futility. Although in the end Rubin introduces social justice to justify denying medical treatment to a patient, she insists that the refusal to provide demanded treatment is justified "if social consensus deems certain treatments to be insufficiently worth pursuing *with respect to other competing social goods*..." Thus even here her ultimate appeal is to social consensus. I believe that a social consensus concerning moral issues can justify a social practice only if it correctly appeals to more fundamental moral reasons. Therefore, any adequate conclusion concerning whether there ought to be a legal liberty of the physician to refuse medical treatment demanded by the patient can be justified only by the kinds of moral arguments I have been discussing, not by attempting to achieve a social consensus.

10.4. Conclusion

I do not pretend to have reported all the arguments for or against a legal liberty of the physician to refuse to provide futile medical treatment demanded by the patient or her family. I have instead selected those that are the most important and have formulated them in the most plausible way. I have then subjected each argument to critical examination in order to assess its relevance and strength. Thus, I have shown that two of the most popular reasons given in favor of such a legal liberty are very weak. Although the argument from physiological futility is often said to be clear and uncontroversial,[98] it is relevant only if one assumes some essential connection between physiological and qualitative futility. Again, the fact that a treatment may be cruel or inhumane, although a good reason to refuse

[96] *Ibid.*, p. 129, italics in original.
[97] *Ibid.*, pp. 135–142.
[98] For example, Griener, *Op. Cit.*, p. 210.

to provide that treatment, bears on the futility of treatment only insofar as the harms it imposes upon the patient serve to render the treatment nonbeneficial on balance.

This leaves four cogent arguments in favor of a legal liberty of the physician to refuse to provide futile medical treatment demanded by the patient or her family. One takes a negative form. It assumes that the law ought not to limit the liberty of the individual without some good reason. But since futile treatment is of no net benefit to the patient and the physician's primary duty is to the patient, there is no good reason to limit the physician's liberty by requiring the physician to provide such treatment. Therefore, the law ought not to impose any such duty upon the physician, which implies that the physician ought to have a legal liberty to refuse to provide futile medical treatment. In addition, there are three positive arguments for the same conclusion. The physician ought to have a legal liberty to refuse to provide futile medical treatment demanded by the patient or her family in order to avoid wasting scarce medical resources, to preserve the physician's professional integrity, and to respect the physician's moral integrity. Together these four arguments are almost conclusive.

What can be argued against this conclusion? Several of the moral rights often said to be violated by a physician's liberty to refuse to provide futile medical treatment demanded by the patient or her family turn out to be irrelevant. These are the patient's right not to be abandoned by her physician, her right to autonomy or self-determination, her right to life, and her right to nondiscrimination or equitable treatment. This is an unexpected conclusion because at first glance each of these fundamental human rights would seem to require the physician to provide even futile treatment if her patient insists upon her doing so.

There is, however, much to be learned from the argument that there are better ways of solving the problems posed by medical futility. There ought to be established processes within health care institutions for resolving disagreements between physicians and patients concerning the value of possible medical treatments. And if an appeal to the law is made, the courts ought to decide medical futility cases on the basis of the standard of medical care in medical malpractice law. This will require that some reasonably objective and concrete definition of medical futility be incorporated into the legal standard of care. Although no such definition is yet available, I can see no reason why judges cannot in the light of good medical practice develop specific criteria to determine when medical treatment would be of no net benefit to the patient in the same way they have come to evaluate benefits and harms to the patient in other areas of medical malpractice law. Thus, the physician's liberty to refuse to provide medical treatment on the basis of his individual judgment that the demanded treatment is futile would be limited by the judgment of the medical profession as a whole. And court cases touching upon issues of great public concern will serve to stimulate the wider public discussion that could lead to a rethinking of the proper place of medical authority in our society. In addition, physicians should have a legal duty to co-operate with those patients who wish to find another physician who is willing and able to provide the treatment they desire. These measures should be sufficient to greatly reduce, if not entirely eliminate, the danger that physicians will abuse their liberty to deny treatment to their patients. I conclude, and trust that my reader will concur, that there ought to be a legal liberty of the physician to refuse to provide futile medical care demanded by the patient or her family.

INDEX OF CASES

INDEX OF NAMES

INDEX OF SUBJECTS